Gladys Horiuchi

Edwin C. ...

Robin Kelley O'Connor

Alistair Robertson

LOUIS LATOUR

Marvin R. Shanken

L F Bouchard

Rodney D. Strong

Rio Roffi...

Stéfan ...

James Trezise

Bo Barrett

David S. Stare

Louis P. Martini

Alan Lewis

Mike Stephens

Ed Shragea

John L. Mers

Claude Taittinger
Jean Hugel
Patrick Campbell
Darrell Corti
Philip di Belardino
Robert de Goulaine
Jack Cakebread
Zelma Long
Georges Duboeuf
Edmund
Hubert Trimbach
Myron S. Nightingale

WINDOWS ON THE WORLD

Complete Wine Course

2009 EDITION

WINDOWS ON THE WORLD
Complete Wine Course

Kevin Zraly

STERLING

New York / London
www.sterlingpublishing.com

This 2009 edition is dedicated to my father, Charles, who taught me my work ethics, values, and common sense. To my uncle, Thomas Cummins, who gave me my first experience working with wine and encouraged me to continue in the restuarant business.

PHOTO & ILLUSTRATION CREDITS

Vineyard near Beaune, Burgundy (pp. ii-iii) © Charles O'Rear/Corbis
Insets p. ii: Wine tasting © Owen Franken/Corbis; *Vineyard near Savigny-les-Beaune, Burgundy* © Georgina Bowater/Corbis; *Château Beycheville, Bordeaux* © Adam Woolfitt/Corbis
Grapes on the vine in Napa, California (pp. xii-1) © Morton Beebe/Corbis
Vineyards and fields in Chablis (pp. 18-19) © Owen Franken/Corbis
Napa Valley vineyards (pp. 44-45) © Charles O'Rear/Corbis
Braune Kupp vineyard, Mosel-Saar-Ruwer, Germany (pp. 82-83) © Charles O'Rear/Corbis
Vineyard at Volnay, Côte de Beaune, Burgundy (pp. 96-97) © Charles O'Rear/Corbis
Cellar at Château Mouton-Rothschild (pp. 116-117) © Charles O'Rear/Corbis
Cabernet Sauvignon vineyard, Napa Valley (pp. 134-135) © Jim Sugar/Corbis
Vineyards, Barolo, Piedmont (pp. 148-149) © John Heseltine/Corbis
Barrels of Port, Dow & Co. warehouse, Oporto, Portugal (pp. 174-175) © Charles O'Rear/Corbis
Grapes, Avignonesi Winery, Tuscany (p. 191) © Bob Krist/Corbis
How Do We Smell? (p. 256), illustration adapted by permission from Tim Jacob, http://www.cf.ac.uk/biosi/staff/jacob

ACKNOWLEDGMENTS

I am indebted to the winemakers, grape growers, and wine friends throughout the world who have contributed their expertise and enthusiasm to this book. The signatures on the endpapers represent some of the people whose help was invaluable to me.

A special thank-you to my family and friends. I can't express enough how much your support has meant to me.

I welcome questions and comments. Please visit my Web site, *windowswineschool.com,* for updates and wine news.

STERLING and the distinctive Sterling logo are registered trademarks of Sterling Publishing Co., Inc.

LIBRARY OF CONGRESS CATALOGING-IN-PUBLICATION DATA AVAILABLE

10 9 8 7 6 5 4 3 2 1

Published by Sterling Publishing Co., Inc.
387 Park Avenue South, New York, NY 10016
© 2008 by Kevin Zraly
Distributed in Canada by Sterling Publishing
c/o Canadian Manda Group, 165 Dufferin Street
Toronto, Ontario, Canada M6K 3H6
Distributed in the United Kingdom by GMC Distribution Services
Castle Place, 166 High Street, Lewes, East Sussex, England BN7 1XU
Distributed in Australia by Capricorn Link (Australia) Pty. Ltd.
P.O. Box 704, Windsor, NSW 2756, Australia

Printed in China.
All rights reserved

STERLING ISBN 978-1-4027-5746-4

For information about custom editions, special sales, premium and corporate purchases, please contact Sterling Special Sales Department at 800-805-5489 or specialsales@sterlingpublishing.com.

In 2008, I am celebrating my 32nd anniversary teaching the Windows on the World Wine School. It all started in 1976 when Windows on the World, the restaurant on top of One World Trade Center, opened. We were a private lunch club, and I was asked to put together a six-week wine course as a club activity. We started off with only twenty people, and I taught the first class and the last class; the other four classes were taught by prominent wine writers. To this day, I have vivid memories of the great wines that we poured for that first class and the "happy" response of the students.

In fact, the club members were so happy with the wine school that they started asking if they could bring friends. By 1979 there were more friends than club members attending the class. The growing interest in wine was becoming obvious. In 1980 we opened our first classes for consumers. To date, more than eighteen thousand people have graduated from the Windows on the World Wine School. Many of my students have found their passion with wine—opening retail stores, restaurants, and wine bars. Some have started importing wine and a few have even started their own wineries.

Since my teenage years, wine has been my passion. It started well before Windows on the World. In 1970, as a nineteen-year-old college student, I got a job as a waiter at a restaurant to pick up some "beer money." The restaurant received a four-star rating from *The New York Times* restaurant critic Craig Claiborne and the owner asked me to take over the bartending duties. (Luckily for me, back then the legal age to drink was eighteen in New York State.) I started studying about beer, distilled spirits, and wine, learning as much as I could. Within six months my interest in wine became an obsession. I quickly switched from Budweiser to Bordeaux. I read every book on wine I could find, and I visited all the wineries in New York. (I couldn't go to California—the drinking age there was twenty-one!)

A local college contacted the restaurant, suggesting a wine-and-cheese course for their adult-education program. So in 1971, at twenty years of age, I taught my first wine class. I continued to teach this class throughout college, and I was beginning to sense that wine appreciation was going to become a major area of interest in America.

I eventually graduated with an education degree and, combining that with my passion for wine, the Wine School for me has become the best of both worlds. After thirty-seven years of teaching about wine I am still fascinated by the subject, intrigued by its complexities, and "thirsty" to learn more with every new vintage. I would like to thank all of my graduates for supporting my passion.

🍷

This book is dedicated to everyone who has a passion for wine, from the grape growers to the winemakers, from the buyers and the sellers to, most important, the wine consumers.

May your glass always be more full than empty.

Contents

Introduction

This book grew out of my involvement with the Wine School.

WHEN I FIRST started the Wine School, I searched everywhere for a simple, straightforward wine guide to use as a textbook. I picked the few that came closest but by the end of each semester I would be looking again. Every book was either too encyclopedic or too scientific. I wasn't the only one who was frustrated: My students were, too. Teaching and learning had stopped being fun; my students were finding the texts too complicated, and I referred to them less and less. So I began developing my own course material, which I kept simple, giving my students only the basic information they needed for understanding and enjoying wine.

I taught from a handwritten outline and photocopied labels, lists, and other information for each class, handing out copies to students for home study and future reference. This proved so popular that many of my students asked me to write a wine book around the materials. In 1981, five years after I started working at Windows on the World and ten years after teaching my first wine class, I set out to write this book.

Full of naive enthusiasm, I started researching publishers, looking for the best and reading as much as I could about the business of publishing. I soon learned that, although *I* thought my idea for an easy-to-understand wine book was brilliant, finding a publisher who agreed was an entirely different matter. At least five of the largest publishers in the industry turned down my book, telling me that a simple guide to wine would never sell. *Rejection* was becoming an all-too-familiar word.

Eventually I was introduced to Sterling Publishing, a company I like to call the "little engine that could." Back then it was a small, family-owned "niche-marketing" publisher, known for its beautiful craft books. (Or, more accurately, *unknown*: Sterling was probably the most successful small publisher you'd never heard of!) Five years ago, this "little" company was purchased by Barnes & Noble, and I am proud and honored to say that *Windows on the World Complete Wine Course*, with more than 2 million copies sold to date, is the best-selling wine book in the United States!

The success of this book is due, in part, to timing: In the 1980s, when this book was first published, many Americans were becoming interested in wine. They wanted to learn about all the aspects of tasting and buying wine—and they were looking for the material to be presented in an informative and entertaining way.

So what has happened in the wine world over the last thirty-seven years?

I've given a great deal of thought to the changes the wine world has undergone over the past thirty-seven years as I tackled the revisions for this edition. As I reread the first edition, I was struck by the enormous growth in the quality and selection of American wines. In 1970 there were fewer than two hundred wineries in the United States; today there are more than four thousand—with wineries and vineyards in all fifty states.

I've also made many changes to the book's organization over the years. For example, On Tasting Wine has been moved up front to the Prelude to Wine: With wine's increased acceptance and popularity, I think understanding the history of wine is far less important than understanding how to taste and appreciate wine. Another change is the inclusion of a 60-Second Wine Expert section in the Prelude to Wine, which I hope readers will find of value; it's been my most effective classroom teaching tool.

I've learned over the past thirty-seven years that familiarity with the world's major grape varieties is essential to an in-depth understanding of wine. I've added two summaries, White Grapes of the World on page 15 and Red Grapes of the World on page 93. These will give you a quick snapshot of and easy reference for the most important grape varieties.

Americans' wine-drinking habits and interests changed as we began consuming more of it. The first edition of this book emphasized French wines, with California wines a close second—although even in the early eighties many "wine experts" were unconvinced of California wines' quality. Also, we did not discuss wines from Australia, Chile, or Argentina; there was no mention of Washington State or Oregon. All are now included.

There is an old adage that goes, "If it ain't broke, don't fix it!" The Prelude chapter is nearly the same as it was in the first edition, as are the chapters on French and German

wines (although the lists of vintages and best producers have been updated).

The biggest news in winemaking over the last twenty years has come from the United States, Italy, Australia, Chile, Argentina, New Zealand, South Africa, and Spain. While I've learned a lot since I began to study wine, my philosophy follows another old saying: "If I had more time, I would have written a shorter book." Despite its title, this book was never meant to be a "complete" wine course, and in fact the hardest part of writing it has been maintaining its brevity. There are many comprehensive wine books out there and I do not hesitate to recommend them. However, my job is to cut to the chase: not presenting everything there is to know about wine—just what you need to know to better appreciate wine.

CURRENT EVENTS IN THE WINE INDUSTRY

- Americans are drinking more and better wine than ever before. Wine consumption in the U.S. reached a record 703 million gallons in 2006—an increase of over 200 million gallons since 1994. In 2007, the U.S. overtook Italy to become the number two consumer of wine—next stop France! By the end of this decade the U.S. will be the world's biggest wine consumer. Wine is becoming more appealing to Americans than either beer or distilled spirits. A recent Gallup Poll revealed that, for the first time, Americans prefer wine to other alcoholic beverages (39% prefer wine, 36% prefer beer, and 21% prefer spirits). The increased number of wine schools throughout the U.S. is just one result of this exciting trend.

- 2007 was the fourteenth consecutive year of growth in U.S. wine consumption, with total wine sales in the U.S. reaching over 300 million cases. For the first time in my years of studying wine, I feel that wine has become an integral part of the American culture.

- There has been a proliferation of wine events across the United States, almost all of which donate some proceeds to charities. In 2007 the Auction Napa Valley raised $9.8 million, and in 2008 the Naples, Florida, Wine Festival raised $14 million.

- More and more, wine lovers are turning to auctions to buy their wine: Worldwide auction sales totaled $33 million in 1994, compared to $243 million in 2007. Auction sales in the U.S. in 2007 were over $208 million.

- French consumption of wine has dropped more than 40 percent since the 1970s and today the French wine industry is in crisis.

- The quality of restaurant wine lists in the United States has never been better—although some restaurants have created "monster" lists for promotional purposes, creating more intimidation for "new" wine drinkers. Others have found the best wines in each category.

- Such major newspapers as *The Wall Street Journal* and *USA Today* have added full-time wine writers to their staffs. (Although *The New York Times* has had the column "Wine Talk" since the early 1970s, most other wine writing was done on a freelance basis.)

- Among all the great wine-producing countries, the biggest change has been in Italy. The quality of Italian wines has improved dramatically—not just in the historically important regions of Piedmont, Tuscany, and Veneto but also in the country's southern regions, such as Sicily, Campania, and Umbria. Second has to be Spain. Not just in Rioja but throughout the country from big, age-worthy reds to pleasant rosés, Spain is producing its best wine ever.

- Chile has become the source of some of the finest Cabernet Sauvignons available, and Argentina has perfected the Malbec grape, using it to make one of the best wines in the world.

- The amount of corked wines, or TCA (see page 304), has disappointed many wine drinkers. The best estimates indicate that 3 to 5 percent of all corked wines have been spoiled by this compound.

- Technology takes on tradition: High-quality wines worldwide are increasingly being bottled with synthetic corks or screw caps, or even sold in "bag-in-the-box" packaging. Expect this trend to continue.

- The 27 members of the European Union are struggling to deal with the complexities of trying to conform each country's wine laws, surplus wine, and government subsidies.

- The Millennial Generation (ages 21–30) has shown the largest percentage increase in wine consumption over GenX (ages 31–42) and the Baby Boomers (ages 43–61).

- Costco is the largest retailer of wine in the United States, with over $750 million in wine sales.

- Through consolidation, the number of wine wholesalers in the United States has declined by 50 percent.

- In 2005, the U.S. Supreme Court ruled on the legality of interstate shipping of wine directly to consumer. As a result, most states are rewriting their wine laws.

- You can now find more books on wine than ever before.

- Beyond organic: Some growers and winemakers are adopting a holistic approach with biodynamic wine (no fertilizer, pesticides, fungicides). There are not yet many biodynamic wines, but the practice is being explored by such elite wineries of the world as Benziger, Chapoutier, Domaine LeFlaive, Zind Humbrecht, Domain Leroy, and Araujo.

- In 2007, there were 27 million visitors to U.S. wineries, totalling $3 billion in sales.

- Women are responsible for 55% of U.S. wine purchases.

- California faces competition: Since 1991, wine imports to the U.S. have doubled, from 12% to 25% in 2007. Most of these imports are inexpensive, often selling at retail for less than 10 dollars.

- All fifty states now have wineries.

- Everyone, it seems, is now involved in wine: Celebrity branding is big, from baseball stars (Tom Seaver) to racecar drivers (Jeff Gordon) to musicians (Bob Dylan). Celebrity wine sales were over $50 million in 2007.

- The link between red wine drinking and health has been widely covered in the media.

- Wine 2.0—wine blogs—now everyone can become a wine critic or ask questions to the critics on erobertparker.com, Wine Spectator Blog Index, or the Wine Enthusiasts unreserved, to name a few.

- Pioneers cashing out: Stag's Leap Wine Cellars and Duckhorn were sold in 2007.

- Wine in a box is becoming more popular—Franzia Winetaps sells 5-liter boxed wine.

- Marketing 101—wine names are becoming more "creative": Red Truck, Pinot Evil, Killer Juice, Dog House, Fat Bastard, Ditka Kick Ass Red, Mad Housewife . . .

- 2007 Worldwide Statistics: 19.6 million acres of vineyards, 75,000 million gallons of wine produced, 63,613 million gallons of wine consumed.

Prelude to Wine

❧

FERMENTATION • GRAPE VARIETIES AND

TYPES OF WINE • WORLD WINE PRODUCTION • THE FIVE BASIC STEPS

OF TASTING WINE • THE 60-SECOND WINE EXPERT

THE FIVE major wine importers into the United States:
1. Italy (32%)
2. Australia (24%)
3. France (20%)
4. Chile (6%)
5. Spain (5%)

ITALY, AUSTRALIA, and France account for about 75% of all imports into the United States.

TWENTY-TWO brand names equal 50% of the total wine market. *Source: Impact Databank*

THE TOP TEN producers of wine in the world:
1. France	6. Australia
2. Italy	7. South Africa
3. Spain	8. Germany
4. United States	9. Chile
5. Argentina	10. Portugal

WINE IS FAT FREE and contains no cholesterol.

WINDOWS ON WINE 2007
World Production - 39 billion bottles
World consumption - 33 billion bottles
Acres of Vineyards - 19.6 million

TOP 3 IMPORTED wines sold in the U.S.:
Yellow Tail (Australia)
Cavit (Italy)
Concha y Toro (Chile)

THE HIGHER the alcohol in a wine, the more body (weight) it will usually have.

ALCOHOL WHEN consumed in moderation will increase HDL (good) and decrease LDL (bad) cholesterol.

A BOTTLE of wine is 86% water.

MORE ACRES of grapes are planted than any other fruit crop in the world!

The Basics

You're in a wine shop looking for that "special" wine to serve at a dinner party. Before you walked in, you had at least an idea of what you wanted, but now, as you scan the shelves, you're overwhelmed. "There are so many wines," you think, "and so many prices." You take a deep breath, boldly pick up a bottle that looks impressive, and buy it. Then you hope your guests will like your selection.

Does this sound a little farfetched? For some of you, yes. Yet the truth is, this is a very common occurrence for the wine beginner, and even someone with intermediate wine knowledge, but it doesn't have to be that way. Wine should be an enjoyable experience. By the time you finish this book, you'll be able to buy with confidence from a retailer or even look in the eyes of a wine steward and ask with no hesitation for the selection of your choice. But first let's start with the basics—the foundation of your wine knowledge. Read carefully, because you'll find this section invaluable as you relate it to the chapters that follow. You may even want to refer back to it occasionally to reinforce what you learn.

For the purpose of this book, wine is the fermented juice of grapes.

What's fermentation?

Fermentation is the process by which the grape juice turns into wine. The simple formula for fermentation is:

$$\text{Sugar} + \text{Yeast} = \text{Alcohol} + \text{Carbon Dioxide (CO}_2\text{)}$$

The fermentation process begins when the grapes are crushed and ends when all of the sugar has been converted to alcohol or the alcohol level has reached around 15 percent, the point at which the alcohol kills off the yeast. Sugar is naturally present in the ripe grape through photosynthesis. Yeast also occurs naturally as the white bloom on the grape skin. However, this natural yeast is not always used in today's winemaking. Laboratory strains of pure yeast have been isolated and may be used in many situations, each strain contributing something unique to the style of the wine. The carbon dioxide dissipates into the air, except in Champagne and other sparkling wines, in which this gas is retained through a special process which we will discuss in Class Eight.

What are the three major types of wine?

Table wine: approximately 8 to 15 percent alcohol

Sparkling wine: approximately 8 to 12 percent alcohol + CO_2

Fortified wine: 17 to 22 percent alcohol

All wine fits into at least one of these categories.

Why do the world's fine wines come only from certain areas?

A combination of factors is at work. The areas with a reputation for fine wines have the right soil and favorable weather conditions, of course. In addition, these areas look at winemaking as an important part of their history and culture.

Is all wine made from the same kind of grape?

No. The major wine grapes come from the species *Vitis vinifera*. In fact, European, North American, Australian, and South American winemakers use the *Vitis vinifera*, which includes many different varieties of grapes—both red and white. However, there are other grapes used for winemaking. The most important native grape species in America is *Vitis labrusca*, which is grown widely in New York State as well as other East Coast and Midwest states. Hybrids, which are also used in modern winemaking, are a cross between *Vitis vinifera* and native American grape species, such as *Vitis labrusca*.

Where are the best locations to plant grapes?

Grapes are agricultural products that require specific growing conditions. Just as you wouldn't try to grow oranges in Maine, you wouldn't try to grow grapes at the North Pole. There are limitations on where vines can be grown. Some of these limitations are: the growing season, the number of days of sunlight, angle of the sun, average temperature, and rainfall. Soil is of primary concern, and adequate drainage is a requisite. The right amount of sun ripens the grapes properly to give them the sugar/acid balance that makes the difference between fair, good, and great wine.

KEVIN ZRALY'S FAVORITE WINE REGIONS

Napa
Sonoma
Bordeaux
Burgundy
Champagne
Rhône Valley
Tuscany
Piedmont
Mosel
Rhine
Rioja
Douro (Port)
Mendoza
Maipo Valley
South Australia

PLANTING OF vineyards for winemaking began more than 8,000 years ago.

A SAMPLING OF THE MAJOR GRAPES

Vitis vinifera
Chardonnay
Cabernet Sauvignon

Vitis labrusca
Concord
Catawba

Hybrids
Seyval Blanc
Baco Noir

VITIS is Latin for vine.
VINUM is Latin for wine.

THE TOP five countries in wine grape acreage worldwide:
  1. Spain
  2. France
  3. Italy
  4. Turkey
  5. United States

Does it matter where grapes are planted?

Yes, it does. Traditionally, many grape varieties produce better wines when planted in certain locations. For example, most red grapes need a longer growing season than do white grapes, so red grapes are usually planted in warmer locations. In colder northern regions—in Germany and northern France, for instance—most vineyards are planted with white grapes. In the warmer regions of Italy, Spain, and Portugal, and in California's Napa Valley, the red grape thrives.

When is the harvest?

Grapes are picked when they reach the proper sugar/acid ratio for the style of wine the vintner wants to produce. Go to a vineyard in June and taste one of the small green grapes. Your mouth will pucker because the grape is so tart and acidic. Return to the same vineyard—even to that same vine—in September or October, and the grapes will taste sweet. All those months of sun have given sugar to the grape as a result of photosynthesis.

What effect does weather have on the grapes?

Weather can interfere with the quality of the harvest, as well as its quantity. In the spring, as vines emerge from dormancy, a sudden frost may stop the flowering, thereby reducing the yields. Even a strong windstorm can affect the grapes adversely at this crucial time. Not enough rain, too much rain, or rain at the wrong time can also wreak havoc.

Rain just before the harvest will swell the grapes with water, diluting the juice and making thin, watery wines. Lack of rain will affect the wines' balance by creating a more powerful and concentrated wine, but will result in a smaller crop. A severe drop in temperature may affect the vines even outside the growing season. Case in point: In New York State the winter of 2003–04 was one of the coldest in fifty years. The result was a major decrease in wine production, with some vineyards losing more than 50 percent of their crop for the 2004 vintage.

What can the vineyard owner do in the case of adverse weather?

A number of countermeasures are available to the grower. Some of these measures are used while the grapes are on the vine; others are part of the winemaking process.

PROBLEM	RESULTS IN	SOME SOLUTIONS
Frost	Reduced yield	Various frost protection methods: wind machines, sprinkler systems, and flaming heaters
Not enough sun	Underripe, green herbal, vegetal character, high acid, low sugar	Chaptalization (the addition of sugar to the must—fresh grape juice—during fermentation)
Too much sun	Overripe, high alcohol, prune character	Amelioration (addition of water)
Too much rain	Thin, watery wines	Move vineyard to a drier climate
Mildew	Rot	Spray with copper sulfate
Drought	Scorched grapes	Irrigate or pray for rain
Phylloxera	Dead vines	Graft vines onto resistant rootstock
High alcohol	Change in the balance of the components	De-alcoholize
High acidity	Sour, tart wine	De-acidify

What is phylloxera?

Phylloxera, a grape louse, is one of the grapevine's worst enemies, because it eventually kills the entire plant. An epidemic infestation in the 1870s came close to destroying all the vineyards of Europe. Luckily, the roots of native American vines are immune to this louse. After this was discovered, all the European vines were pulled up and grafted onto phylloxera-resistant American rootstocks.

WALNUTS AND tea also contain tannin.

BESIDES TANNIN, red wine contains resveratrol, which in medical studies has been associated with anticancer properties.

"And Noah began to be a husbandman and he planted a vineyard, and he drank of the vine."
                    GENESIS 9:20–21

KING TUTANKHAMEN, who died in 1327 B.C., apparently preferred the taste of red wine, according to scientists who found residues of red wine compounds in ancient Egyptian jars found in his tomb.

THE FIRST known reference to a specific vintage was made by Roman scientist Pliny the Elder, who rated the wines of 121 B.C. "of the highest excellence."

2005 WAS a great vintage year in every major wine region on earth!

THREE MAJOR wine collectibles that will age more than ten years:
  1. Great châteaux of Bordeaux
  2. Best producers of California Cabernet Sauvignon
  3. Finest producers of vintage Port

"The truth of wine aging is that it is unknown, unstudied, poorly understood, and poorly predicted!"
              —ZELMA LONG, *California winemaker*

Can white wine be made from red grapes?

Yes. The color of wine comes entirely from the grape skins. By removing the skins immediately after picking, no color is imparted to the wine, and it will be white. In the Champagne region of France, a large percentage of the grapes grown are red, yet most of the resulting wine is white. California's White Zinfandel is made from red Zinfandel grapes.

What is tannin, and is it desirable in wine?

Tannin is a natural substance that comes from the skins, stems, and pips of the grapes, and even from the wooden barrels in which many are aged. It is a natural preservative; without it, certain wines wouldn't continue to improve in the bottle. In young wines, tannin can be very astringent and make the wine taste bitter. Generally, red wines have a higher level of tannin than whites, because red grapes are usually left to ferment with their skins.

Is acidity desirable in wine?

All wine will have a certain amount of acidity. Generally, white wines have more perceived acidity than reds, though winemakers try to have a balance of fruit and acid. An overly acidic wine is also described as tart or sour. Acidity is a very important component in the aging of wines.

What is meant by "vintage"? Why is one year considered better than another?

A vintage indicates the year the grapes were harvested, so every year is a vintage year. A vintage chart reflects the weather conditions for various years. Better weather usually results in a better rating for the vintage, and therefore a higher likelihood that the wine will age well.

Are all wines meant to be aged?

No. It's a common misconception that all wines improve with age. In fact, more than 90 percent of all the wines made in the world should be consumed within one year, and less than 1 percent of the world's wines should be aged for more than five years. Wines change with age. Some get better, but most do not. The good news is that the 1 percent represents more than 35 million bottles of wine every vintage.

What makes a wine last more than five years?

The color and the grape: Red wines, because of their tannin content, will generally age longer than whites. And certain red grapes, such as Cabernet Sauvignon, tend to have more tannin than, say, Pinot Noir.

The vintage: The better the weather conditions in one year, the more likely the wines from that vintage will have a better balance of fruits, acids, and tannins, and therefore have the potential to age longer.

Where the wine comes from: Certain vineyards have optimum conditions for growing grapes, including such factors as soil, weather, drainage, and slope of the land. All of this contributes to producing a great wine that will need time to age.

How the wine was made (vinification): The longer the wine remains in contact with its skins during fermentation (maceration), and if it is fermented and/or aged in oak, the more of the natural preservative tannin it will have, which can help it age longer. These are just two examples of how winemaking can affect the aging of wine.

Wine storage conditions: Even the best-made wines in the world will not age well if they are improperly stored. (See page 295.)

How is wine production regulated worldwide?

Each major wine-producing country has government-sponsored control agencies and laws that regulate all aspects of wine production and set certain minimum standards that must be observed. Here are some examples:

France: Appellation d'Origine Contrôlée (AOC)

Italy: Denominazione di Origine Controllata (DOC)

United States: Alcohol and Tobacco Tax and Trade Bureau

Germany: Ministry of Agriculture

Spain: Denominación de Origen (DO)

"I like to think about the life of wine, how it is a living thing. I like to think about what was going on the year the grapes were growing. How the sun was shining, if it rained. I like to think about all the people who tended and picked the grapes. And if it's an old wine, how many of them must be dead by now. I like how wine continues to evolve, like if I opened a bottle of wine today it would taste different than if I'd opened it on any other day, because a bottle of wine is actually alive. And it's constantly evolving and gaining complexity, that is until it peaks, like your '61. And then it begins its steady inevitable decline."
—MAYA, *from the movie* Sideways *(2004)*

5 Bottles of wine produced annually from one grapevine

240 Bottles of wine in a barrel

720 Bottles of wine from a ton of grapes

5,500 Bottles of wine produced annually from an acre of grapevines

Source: Napa Valley Vintners

THERE ARE more than seventy wine-producing countries in the world.

ACCORDING TO *Wines & Vines,* the value of wine sold worldwide is now more than $100 billion.

TOP 10 MOST POWERFUL WINE BRANDS OF 2007

1. Moet & Chandon
2. Gallo
3. Hardy's
4. Concha y Toro
5. Veuve Cliquot
6. Robert Mondavi
7. Yellow Tail
8. Freixenet
9. Jacob's Creek
10. Lindeman's

Source: Intangible Business

IF YOU can see through a red wine, it's generally ready to drink!

AS WHITE wines age, they gain color. Red wines, on the other hand, lose color as they age.

TYPES OF TASTINGS

HORIZONTAL	Tasting wines from the same vintage
VERTICAL	Comparing wines from different vintages
BLIND	The taster does not have any information about the wines
SEMI-BLIND	The taster knows only the style of wine (grape) or where it comes from

ON TASTING WINE

You can read all the books (and there are plenty) written on wine to become more knowledgeable on the subject, but the best way to truly enhance your understanding of wine is to taste as many wines as possible. Reading covers the more academic side of wine, while tasting is more enjoyable and practical. A little of each will do you the most good.

The following are the necessary steps for tasting wine. You may wish to follow them with a glass of wine in hand.

Wine tasting can be broken down into five basic steps: Color, Swirl, Smell, Taste, and Savor.

Color

The best way to get an idea of a wine's color is to get a white background—a napkin or tablecloth—and hold the glass of wine on an angle in front of it. The range of colors that you may see depends, of course, on whether you're tasting a white or red wine. Here are the colors for both, beginning with the youngest wine and moving to an older wine:

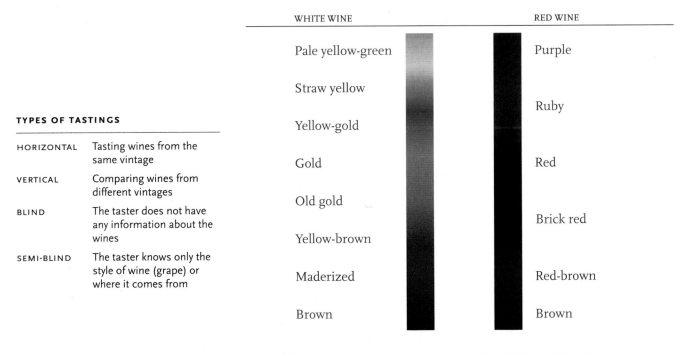

WHITE WINE		RED WINE
Pale yellow-green		Purple
Straw yellow		
Yellow-gold		Ruby
Gold		Red
Old gold		
Yellow-brown		Brick red
Maderized		Red-brown
Brown		Brown

Color tells you a lot about the wine. Since we start with the white wines, let's consider three reasons why a white wine may have more color:

1. It's older.
2. Different grape varieties give different color. (For example, Chardonnay usually gives off a deeper color than does Sauvignon Blanc.)
3. The wine was aged in wood.

In class, I always begin by asking my students what color the wine is. It's not unusual to hear that some believe that the wine is pale yellow-green, while others say it's gold. Everyone begins with the same wine, but color perceptions vary. There are no right or wrong answers, because perception is subjective. So you can imagine what happens when we actually taste the wine!

Swirl

Why do we swirl wine? To allow oxygen to get into the wine. Swirling releases the esters, ethers, and aldehydes that combine with oxygen to yield a wine's bouquet. In other words, swirling aerates the wine and releases more of the bouquet and aroma.

Smell

This is the most important part of wine tasting. You can perceive just four tastes—sweet, sour, bitter, and salty—but the average person can identify more than two thousand different scents, and wine has more than two hundred of its own. Now that you've swirled the wine and released the bouquet, I want you to smell the wine at least three times. You may find that the third smell will give you more information than the first smell did. What does the wine smell like? What type of nose does it have? Smell is the most important step in the tasting process and most people simply don't spend enough time on it.

Pinpointing the nose of the wine helps you to identify certain characteristics. The problem here is that many people in class want me to tell them what the wine smells like. Since I prefer not to use subjective words, I may say that the wine smells like a French white Burgundy. Still, I find that this doesn't satisfy the majority of the class. They want to know more. I ask these people to describe what steak and onions smell like. They answer, "Like steak and onions." See what I mean?

The best way to learn what your own preferences are for styles of wine is to "memorize" the smell of the individual grape varieties. For white, just try to

SOME WINE experts say that if you put your hand over the glass while you swirl, you will get a better bouquet and aroma.

BOUQUET IS the total smell of the wine. AROMA is the smell of the grapes. "NOSE" IS a word that wine tasters use to describe the bouquet and aroma of the wine.

THIS JUST in: It is now known that each nostril can detect different smells.

THE OLDEST part of the human brain is the olfactory region.

THE 2004 Nobel prize for medicine was awarded to two scientists for their research on the olfactory system and the discovery that there are more than 10,000 different smells!

ONE OF THE most difficult challenges in life is to match a smell or a taste with a word that describes it.

SOME CLASSIC WINE DESCRIPTORS

Zinfandel	spiciness
Cabernet Sauvignon	chocolate
Old Bordeaux	wet fallen leaves
Old Burgundy	gamey, mushrooms
Rhône	black pepper
Pouilly-Fumé or Sancerre	gunflint
Chablis	mineral
White Burgundy	chalky
Chardonnay	buttery

WHAT KIND of wine do I like? I like my wine bright, rich, mature, developed, seductive, and with nice legs!

OXYGEN CAN be the best friend of a wine, but it can also be its worst enemy. A little oxygen helps release the smell of the wine (as with swirling), but prolonged exposure can be harmful, especially to older wines.

EVERY WINE contains a certain amount of sulfites. They are a natural by-product of fermentation.

EACH PERSON has a different threshold for sulfur dioxide, and although most people do not have an adverse reaction, it can be a problem for individuals with asthma. To protect those who are prone to bad reactions to sulfites, federal law requires winemakers to label their wines with the warning that the wine contains sulfites.

memorize the three major grape varieties: Chardonnay, Sauvignon Blanc, and Riesling. Keep smelling them, and smelling them, and smelling them until you can identify the differences, one from the other. For the reds it's a little more difficult, but you still can take three major grape varieties: Pinot Noir, Merlot, and Cabernet Sauvignon. Try to memorize those smells without using flowery words, and you'll understand what I'm talking about.

For those in the Wine School who remain unconvinced, I hand out a list of five hundred different words commonly used to describe wine. Here is a small excerpt:

acetic	character	legs	seductive
aftertaste	corky	light	short
aroma	delicate	maderized	soft
astringent	developed	mature	stalky
austere	earthy	metallic	sulfury
baked-burnt	finish	moldy	tart
balanced	flat	nose	thin
big-full-heavy	fresh	nutty	tired
bitter	grapey	off	vanilla
body	green	oxidized	woody
bouquet	hard	pétillant	yeasty
bright	hot	rich	young

You're also more likely to recognize some of the defects of a wine through your sense of smell.

Following is a list of some of the negative smells in wine:

SMELL	WHY
Vinegar	Too much acetic acid in wine
Sherry*	Oxidation
Dank, wet, moldy, cellar smell	Wine absorbs the taste of a defective cork (referred to as "corked wine")
Sulfur (burnt matches)	Too much sulfur dioxide

* Authentic Sherry, from Spain, is intentionally made through controlled oxidation.

All wines contain some sulfur dioxide since it is a by-product of fermentation. Sulfur dioxide is also used in many ways in winemaking. It kills bacteria in wine, prevents unwanted fermentation, and acts as a preservative. It sometimes causes a burning and itching sensation in your nose.

Taste

To many people, tasting wine means taking a sip and swallowing immediately. To me, this isn't tasting. Tasting is something you do with your taste buds. You have taste buds all over your mouth—on both sides of the tongue, underneath, on the tip, and extending to the back of your throat. If you do what many people do, you take a gulp of wine and bypass all of those important taste buds. When I taste wine I leave it in my mouth for three to five seconds before swallowing. The wine warms up, sending more of the bouquet and aroma up through the nasal passage then on to the olfactory bulb, and then to the limbic system of the brain. Remember, 90 percent of taste is smell.

What should you think about when tasting wine?

Be aware of the most important sensations of taste and your own personal thresholds to those tastes. Also, pay attention to where they occur on your tongue and in your mouth. As I mentioned earlier, you can perceive just four tastes: sweet, sour, bitter, and salty (but there's no salt in wine, so we're down to three). Bitterness in wine is usually created by high alcohol and high tannin. Sweetness occurs only in wines that have some residual sugar left over after fermentation. Sour (sometimes called "tart") indicates the acidity in wine.

Sweetness: The highest threshold is on the tip of the tongue. If there's any sweetness in a wine whatsoever, you'll get it right away.

Acidity: Found at the sides of the tongue, the cheek area, and the back of the throat. White wines and some lighter-style red wines usually contain a higher degree of acidity.

Bitterness: Tasted on the back of the tongue.

Tannin: The sensation of tannin begins in the middle of the tongue. Tannin frequently exists in red wines or white wines aged in wood. When the wines are too young, tannin dries the palate to excess. If there's a lot of tannin in the wine, it can actually coat your whole mouth, blocking the fruit. Remember, tannin is not a taste: It is a tactile sensation.

Fruit and varietal characteristics: These are not tastes, but smells. The weight of the fruit (the "body") will be felt in the middle of the tongue.

Aftertaste: The overall taste and balance of the components of the wine that lingers in your mouth. How long does the balance last? Usually a sign of a high-quality wine is a long, pleasing aftertaste. The taste of many of the great wines lasts anywhere from one to three minutes, with all their components in harmony.

THE AVERAGE person has 5,000 taste buds.

TASTING WINE is confirming what the color and smell are telling you.

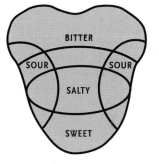

THERE IS now evidence that people may perceive five tastes: sweet, sour, bitter, salty, and possibly umami—aka MSG.

OTHER SENSATIONS associated with wine include numbing, tingling, drying, cooling, warming, and coating.

TANNIN: Think gritty.
BITTER: Think endive or arugula.

Savor

After you've had a chance to taste the wine, sit back for a few moments and savor it. Think about what you just experienced, and ask yourself the following questions to help focus your impressions.

WINE TEXTURES:
   Light—skim milk
   Medium—whole milk
   Full—heavy cream

- Was the wine light, medium, or full-bodied?

- For a white wine: How was the acidity? Very little, just right, or too much?

- For a red wine: Is the tannin in the wine too strong or astringent? Does it blend with the fruit or overpower it?

- What is the strongest component (residual sugar, fruit, acid, tannin)?

- How long did the balance of the components last (ten seconds, sixty seconds, etc.)?

- Is the wine ready to drink? Or does it need more time to age? Or is it past its prime?

- What kind of food would you enjoy with the wine?

- To your taste, is the wine worth the price?

- This brings us to the most important point. The first thing you should consider after you've tasted a wine is whether or not you like it. Is it your style?

"The key to great wine is balance, and it is the sum of the different parts that make a wine not only delicious but complete and fascinating as well as worthy of aging."
   —FIONA MORRISON, M.W.

"One not only drinks wine, one smells it, observes it, tastes it, sips it, and—one talks about it."
   —KING EDWARD VII OF ENGLAND

"A wine goes in my mouth, and I just see it. I see it in three dimensions. The textures. The flavors. The smells. They just jump out at me. I can taste with a hundred screaming kids in a room. When I put my nose in a glass, it's like tunnel vision. I move into another world, where everything around me is just gone, and every bit of mental energy is focused on that wine."
   —ROBERT M. PARKER JR., *author and wine critic, in* The Atlantic Monthly

You can compare tasting wine to browsing in an art gallery. You wander from room to room looking at the paintings. Your first impression tells whether or not you like something. Once you decide you like a piece of art, you want to know more: Who was the artist? What is the history behind the work? How was it done? And so it is with wine. Usually, once oenophiles (wine aficionados) discover a wine that they like, they want to learn everything about it: the winemaker; the grapes; exactly where the vines were planted; the blend, if any; and the history behind the wine.

How do you know if a wine is good or not?

The definition of a good wine is one that you enjoy. I cannot emphasize this enough. Trust your own palate and do not let others dictate taste to you!

When is a wine ready to drink?

This is one of the most frequently asked questions in my Wine School. The answer is very simple: when all components of the wine are in balance to your particular taste.

The 60-Second Wine Expert

Over the last few years I have insisted that my students spend one minute in silence after they swallow the wine. I use a "60-second wine expert" tasting sheet in my classes for students to record their impressions. The minute is divided into four sections: 0 to 15 seconds, 15 to 30 seconds, 30 to 45 seconds, and the final 45 to 60 seconds. Try this with your next glass of wine.

Please note that the first taste of wine is a shock to your taste buds. This is due to the alcohol content, acidity, and sometimes the tannin in the wine. The higher the alcohol or acidity, the more of a shock. For the first wine in any tasting, it is probably best to take a sip and swirl it around in your mouth, but don't evaluate it. Wait another thirty seconds, try it again, and then begin the 60-second wine expert.

0 to 15 seconds: If there is any residual sugar/sweetness in the wine, I will experience it now. If there is no sweetness in the wine, the acidity is usually at its strongest sensation in the first fifteen seconds. I am also looking for the fruit level of the wine and its balance with the acidity or sweetness.

15 to 30 seconds: After the sweetness or acidity, I am looking for great fruit sensation. After all, that is what I am paying for! By the time I reach thirty seconds, I am hoping for balance of all the components. By this time, I can identify the weight of the wine. Is it light, medium, or full-bodied? I am now starting to think about what kind of food I can pair with this wine (see page 268).

30 to 45 seconds: At this point I am beginning to formulate my opinion of the wine, whether I like it or not. Not all wines need sixty seconds of thought. Lighter-style wines, such as Rieslings, will usually show their best at this point. The fruit, acid, and sweetness of a great German Riesling should be in perfect harmony from this point on. For quality red and white wines, acidity—which is a very strong component (especially in the first thirty seconds)—should now be in balance with the fruit of the wine.

45 to 60 seconds: Very often wine writers use the term "length" to describe how long the components, balance, and flavor continue in the mouth. I concentrate on the length of the wine in these last fifteen seconds.

"Great wine is about nuance, surprise, subtlety, expression, qualities that keep you coming back for another taste. Rejecting a wine because it is not big enough is like rejecting a book because it is not long enough, or a piece of music because it is not loud enough."
—KERMIT LYNCH,
Adventures on the Wine Route

"Wine makes daily living easier, less hurried, with fewer tensions and more tolerance."
—BENJAMIN FRANKLIN

Step One: Look at the color of the wine.
Step Two: Smell the wine three times.
Step Three: Put the wine in your mouth and leave it there for three to five seconds.
Step Four: Swallow the wine.
Step Five: Wait and concentrate on the wine for 60 seconds before discussing it.

THE CELEBRATION OF WINE AND LIFE: THE TOAST

To complete the five senses (sight, hearing, smell, taste, and touch), don't forget to toast your family and friends with the clinking of the glasses. This tradition started in ancient times when the Greeks, afraid of being poisoned by their enemies, shared a little of their wine with one another. If someone had added something to the wine, it would be a short evening for everyone! The clinking of the glasses *also* is said to drive away the "bad spirits" that might exist and cause the next-day hangover!

*"Wine is the best of all beverages . . . because
it is purer than water, safer than milk, plainer
than soft drinks, gentler than spirits, nimbler
than beer, and ever so much more pleasant to
the educated senses of sight, smell, and taste
than any of the drinkable liquids known to us."*
—ANDRE L. SIMON, *author and
founder of the Wine & Food Society*

WHAT MAKES A GREAT WINE GREAT?

Varietal character
Balance of components
Complexity
Sense of place
Emotional response

In big, full-bodied red wines from Bordeaux and the Rhône Valley, Cabernets from California, Barolos and Barbarescos from Italy, and even some full-bodied Chardonnays, I am concentrating on the level of tannin in the wine. Just as the acidity and fruit balance are my major concerns in the first thirty seconds, it is now the tannin and fruit balance I am looking for in the last thirty seconds. If the fruit, tannin, and acid are all in balance at sixty seconds, then I feel that the wine is probably ready to drink. Does the tannin overpower the fruit? If it does at the sixty-second mark, I will then begin to question whether I should drink the wine now or put it away for more aging.

It is extremely important to me that if you want to learn the true taste of the wine, you take at least one minute to concentrate on all of its components. In my classes it is amazing to see more than a hundred students silently taking one minute to analyze a wine. Some close their eyes, some bow their heads in deep thought, others write notes.

One final point: Sixty seconds to me is the minimum time to wait before making a decision about a wine. Many great wines continue to show balance well past 120 seconds. The best wine I ever tasted lasted more than three minutes—that's three minutes of perfect balance of all components!

FOR FURTHER READING

I recommend Michael Broadbent's *Pocket Guide to Wine Tasting*; Jancis Robinson's *Vintage Timecharts*; and Alan Young's *Making Sense of Wine*.

White Grapes of the World

NOW THAT YOU KNOW THE BASICS of how wine is made and how to taste it, you're almost ready to begin the first three classes on white wines.

Before you do, simplify your journey by letting me answer the question most frequently asked by my wine students on what will help them most in learning about wine. The main thing is to understand the major grape varieties and where they are grown in the world.

The purpose of this book is not to overwhelm you with information about every grape under the sun. My job as a wine educator is to try to narrow down this over-abundance of data. So let's start off with the three major grapes you need to know to understand white wine. More than 90 percent of all quality white wine is made from these three grapes. They are listed here in order from the lightest style to the fullest:

Riesling   Sauvignon Blanc   Chardonnay

This is not to say that world-class white wine comes from only these grapes, but knowing these three is a good start.

One of the first things I show my students in Class One is a list indicating where these three grape varieties grow best. It looks something like this:

GRAPES	WHERE THEY GROW BEST
Riesling	Germany; Alsace, France; New York State; Washington State
Sauvignon Blanc	Bordeaux, France; Loire Valley, France; New Zealand; California (Fumé Blanc)
Chardonnay	Burgundy, France; Champagne, France; California; Australia

There are world-class Rieslings, Sauvignon Blancs, and Chardonnays made in other countries, but in general the above regions specialize in wines made from these grapes.

COMMON AROMAS

Riesling	*Sauvignon Blanc*	*Chardonnay*
Fruity	Grapefruit	Green apple, Butter, Citrus
Lychee nut	Grass, Herbs	Grapefruit, Melon, Oak
Sweet	Cat pee, Green olive	Pineapple, Toast, Vanilla

MORE THAN 50 major white wine grape varieties are grown throughout the world.

OTHER WHITE grapes and regions you may wish to explore:

GRAPES	WHERE THEY GROW BEST
Albariño	Spain
Chenin Blanc	Loire Valley, France; California
Gewürztraminer, Pinot Blanc, Pinot Gris	Alsace, France
Pinot Grigio (aka Pinot Gris)	Italy; California; Oregon
Sémillon	Bordeaux (Sauternes); Australia
Viognier	Rhône Valley, France; California
Grüner Veltliner	Austria

NEW WORLD VS. OLD WORLD

Wines from the United States, Australia, Chile, Argentina, New Zealand, and South Africa usually list the grape variety on the label. French, Italian, and Spanish wines usually list the region, village, or vineyard where the wine was made—but not the grape.

Questions for Prelude

The White Wines of France

UNDERSTANDING FRENCH WINE • ALSACE • LOIRE VALLEY •

WHITE WINES OF BORDEAUX • GRAVES • SAUTERNES/BARSAC •

WHITE WINES OF BURGUNDY • CHABLIS • CÔTE DE BEAUNE •

CÔTE CHÂLONNAISE • MÂCONNAIS

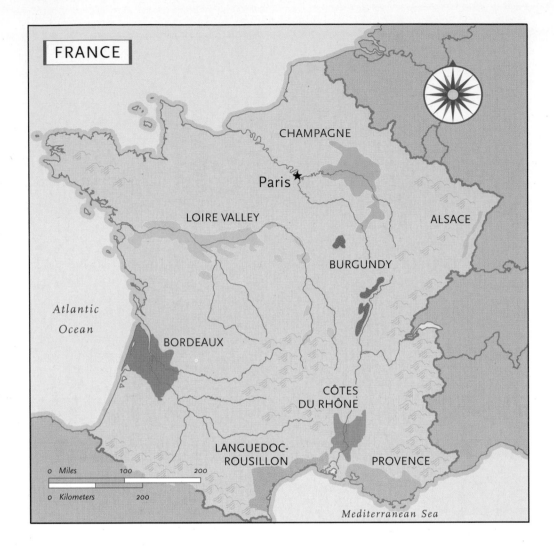

FRANCE

CHAMPAGNE

Paris ★

LOIRE VALLEY

ALSACE

BURGUNDY

Atlantic
Ocean

BORDEAUX

CÔTES
DU RHÔNE

LANGUEDOC-
ROUSILLON

PROVENCE

o Miles 100 200

o Kilometers 200

Mediterranean Sea

Understanding French Wine

France is the #1 producer of wines in the world.

BEFORE WE BEGIN OUR FIRST CLASS, "The White Wines of France," I think you should know a few important points about all French wines. Take a look at a map of France to get familiar with the main wine-producing areas. As we progress, you'll understand why geography is so important.

Here's a quick rundown of which areas produce which kinds of wine:

WINE REGIONS	MAJOR GRAPES
Champagne—sparkling wine	Pinot Noir, Chardonnay
Loire Valley—mostly white	Sauvignon Blanc, Chenin Blanc, Cabernet Franc
Alsace—mostly white	Riesling, Gewürztraminer
Burgundy—red and white	Pinot Noir, Gamay, Chardonnay
Bordeaux—red and white	Sauvignon Blanc, Sémillon, Merlot, Cabernet Sauvignon, Cabernet Franc
Côtes du Rhône—mostly red	Syrah, Grenache
Languedoc-Roussillon—red and white	Carignan, Grenache, Syrah, Cinsault, Mouvedre
Provence—red, white, and rosé	Grenache, Syrah

FRENCH CONTROL LAWS

Established in the 1930s, the Appellation d'Origine Contrôlée (AOC) laws set minimum requirements for each wine-producing area in France. These laws also can help you decipher French wine labels, since the AOC controls the following:

	EXAMPLE	EXAMPLE
1. *Geographic origin*	*Chablis*	*Pommard*
2. *Grape variety: Which grapes can be planted where.*	*Chardonnay*	*Pinot Noir*
3. *Minimum alcohol content: This varies depending upon the particular area where the grapes are grown.*	*10%*	*10.5%*
4. *Vine-growing practices: For example, a vintner can produce only so much wine per acre.*	*40 hectoliters/ acre*	*35 hectoliters/ acre*

IN THE SOUTHERN French region of Provence, look for these producers:
  Domaine Tempier
  Château Routas

FAMOUS NON-AOC French wines that are available in the United States include: Moreau, Boucheron, Chantefleur, and René Junot.

ONLY 35% OF all French wines are worthy of the AOC designation.

THERE ARE more than 465 AOC French wines.

HECTARE—metric measure
1 hectare = 2.471 acres

HECTOLITER—metric measure
1 hectoliter = 26.42 U.S. gallons

TOP FRENCH WINE BRANDS

  Georges Duboeuf – Beaujolais
  Louis Jadot – Burgundy
  Fat Bastard
  B & G
  Red Bicyclette

THE REGION most active in the production of Vin de Pays varietal wines is the Midi, also called Languedoc-Roussillon, in southwest France. Called in the past the "wine lake" because of the vast quantities of anonymous wine made there, the Languedoc has more than 700,000 acres of vineyards, and produces more than 200 million cases a year, about a third of the total French crop.

WHY WOULD Georges Duboeuf, Louis Latour, and many other famous winemakers start wineries in the Midi? For one thing, Midi vineyard land is much cheaper than land in places such as Burgundy or Bordeaux, so the winemakers can produce moderately priced wines and still get a good return on their investment.

CHAMPAGNE IS another major white wine producer, but that's a chapter in itself.

Anyone who is interested in wine is bound to encounter French wine at one time or another. Why? Because of thousands of years of history and winemaking tradition, because of the great diversity and variety of wines from the many different regions, and because French wines have the reputation for being among the best in the world. There's a reason for this, and it goes back to quality control.

French winemaking is regulated by strict government laws that are set up by the **Appellation d'Origine Contrôlée**. If you don't want to say "Appellation d'Origine Contrôlée" all the time, you can simply say "AOC." This is the first of many wine lingo abbreviations you'll learn in this book.

Vins de Pays: This is a category that's growing in importance. A 1979 French legal decision liberalized the rules for this category, permitting the use of nontraditional grapes in certain regions and even allowing vintners to label wines with the varietal rather than the regional name. For exporters to the American market, where consumers are becoming accustomed to buying wines by grape variety—Cabernet Sauvignon or Chardonnay, for example—this change makes their wines much easier to sell.

Vins de Table: These are ordinary table wines and represent almost 35 percent of all wines produced in France.

Most French wine is meant to be consumed as a simple beverage. Many of the *vins de table* are marketed under proprietary names and are the French equivalent of California jug wines. Don't be surprised if you go into a grocery store in France to buy wine and find it in a plastic wine container with no label on it! You can see the color through the plastic—red, white, or rosé—but the only marking on the container is the alcohol content, ranging from 9 to 14 percent. You choose your wine depending on how sharp you need to be for the rest of the day!

When you buy wines, keep these distinctions in mind, because there's a difference not only in quality but also in price.

What are the four major white wine–producing regions of France?

ALSACE   **LOIRE VALLEY**   **BORDEAUX**   **BURGUNDY**

Let's start with Alsace and the Loire Valley, because these are the two French regions that specialize in white wines. As you can see from the map at the beginning of this chapter, Alsace, the Loire Valley, and Chablis (a white wine–producing region of Burgundy) have one thing in common: They're all located in the northern region of France. These areas produce white wines predominantly, because of the shorter growing season and cooler climate which are best suited for growing white grapes.

ALSACE

I often find that people are confused about the difference between wines from Alsace and those from Germany. Why do you suppose this is?

First of all, your confusion could be justified since both wines are sold in tall bottles with tapering necks. Just to confuse you further, Alsace and Germany grow the same grape varieties. But when you think of Riesling, what are your associations? You'll probably answer "Germany" and "sweetness." That's a very typical response, and that's because the German winemaker adds a small amount of naturally sweet unfermented grape juice back into the wine to create the distinctive German Riesling. The winemaker from Alsace ferments every bit of the sugar in the grapes, which is why 99 percent of all Alsace wines are totally dry.

Another fundamental difference between wine from Alsace and wine from Germany is the alcohol content. Wine from Alsace has 11 to 12 percent alcohol, while most German wine has a mere 8 to 9 percent.

What are the white grapes grown in Alsace?

The four grapes you should know are:

Riesling: accounts for 23 percent
Gewürztraminer: accounts for 19 percent
Pinot Blanc: accounts for 20 percent
Pinot Gris: accounts for 7 percent

FROM 1871 TO 1919, Alsace was part of Germany.

ALL WINES produced in Alsace are AOC-designated and represent nearly 20% of all AOC white wines in France.

ALSACE PRODUCES 8% of its red wines from the Pinot Noir grape. These generally are consumed in the region and are rarely exported.

WINE LABELING in Alsace is different from the other French regions administered by the AOC, because Alsace is the only region that labels its wine by varietal. All Alsace wines that include the name of the grape on the label must be made entirely from that grape.

GREAT SWEET (LATE HARVEST) WINES FROM ALSACE

Vendange Tardive
Sélection de Grains Nobles

What types of wine are produced in Alsace?

As mentioned earlier, virtually all the Alsace wines are dry. Riesling is the major grape planted in Alsace, and it is responsible for the highest-quality wines of the region. Alsace is also known for its Gewürztraminer, which is in a class by itself. Most people either love it or hate it, because Gewürztraminer has a very distinctive style. *Gewürz* is the German word for "spice," which aptly describes the wine.

Pinot Blanc and Pinot Gris are becoming increasingly popular among the growers of Alsace.

How should I select an Alsace wine?

Two factors are important in choosing a wine from Alsace: the grape variety and the reputation and style of the shipper. Some of the most reliable shippers are:

DOMAINE MARCEL DEISS   F. E. TRIMBACH

DOMAINE WEINBACH   HUGEL & FILS

DOMAINE ZIND-HUMBRECHT   LÉON BEYER

DOPFF "AU MOULIN"

Why are the shippers so important?

The majority of the landholders in Alsace don't grow enough grapes to make it economically feasible to produce and market their own wine. Instead, they sell their grapes to a shipper who produces, bottles, and markets the wine under his own name. The art of making high-quality wine lies in the selection of grapes made by each shipper.

What are the different quality levels of Alsace wines?

Quality of Alsace wines is determined by the shipper's reputation rather than any labeling on the bottle. That said, the vast majority of any given Alsace wine is the shipper's varietal. A very small percentage is labeled with a specific vineyard's name, especially in the appellation "Alsace Grand Cru." Some wines are labeled "Réserve" or "Réserve Personelle," terms that are not legally defined.

Should I lay down my Alsace wines for long aging?

In general, most Alsace wines are made to be consumed young—that is, one to five years after they're bottled. As in any fine-wine geographic area, in

Alsace there is a small percentage of great wines produced that may be aged for ten years or more.

What are the recent trends in Alsace wines?

What I've learned over the last twenty years is that the more I drink Alsace wines, the more I like them. They're fresh, they're "clean," they're easy to drink, they're very compatible with food.

I think Riesling is still the best grape, but Pinot Blanc, which is lighter in style, is a perfect apéritif wine at a very good price.

Most Alsace wines are very affordable, of good quality, and are available in most markets.

BEST BETS FOR RECENT VINTAGES FROM ALSACE

2000* 2001* 2002* 2003 2004 2005*

*Note: * signifies exceptional vintage*

FOR FURTHER READING

I recommend *Alsace* by S. F. Hallgarten and *Alsace* by Pamela Van Dyke Price.

ALSACE IS also known for its fruit brandies, or eaux-de-vie:

Fraise: Strawberries
Framboise: Raspberries
Kirsch: Cherries
Mirabelle: Yellow plums
Poire: Pears

FOR THE TOURISTS

Visit the beautiful wine village of Riquewihr, whose buildings date from the 15th and 16th centuries.

WINE AND FOOD

During a visit to Alsace, I spoke with two of the region's best-known producers to find out which types of food they enjoy with Alsace wines. Here's what they prefer:

ÉTIENNE HUGEL: "Alsace wines are not only suited to classic Alsace and other French dishes. For instance, I adore Riesling with such raw fish specialties as Japanese sushi and sashimi, while our Gewürztraminer is delicious with smoked salmon and brilliant with Chinese, Thai, and Indonesian food."

Mr. Hugel describes Pinot Blanc as "round, soft, not aggressive . . . an all-purpose wine . . . can be used as an apéritif, with all kinds of pâté and charcuterie, and also with hamburgers. Perfect for brunch—not too sweet or flowery."

HUBERT TRIMBACH: "Riesling with fish—blue trout with a light sauce." He recommends Gewürztraminer as an apéritif, or with foie gras or any pâté at meal's end; with Muenster cheese, or a stronger cheese such as Roquefort.

ALSACE
APPELLATION ALSACE CONTRÔLÉE

HVH
DEPUIS 1839

APPELLATION ALSACE CONTRÔLÉE
GEWÜRZTRAMINER
"HUGEL"®
ALC. 13% VOL.   750 ml e
MISE EN BOUTEILLE PAR HUGEL ET FILS · RIQUEWIHR · ALSACE · FRANCE
PRODUCE OF FRANCE   BOTTLED IN FRANCE

FOR THE FOODIES

Alsace boasts three Michelin three-star restaurants within a 30-mile radius.

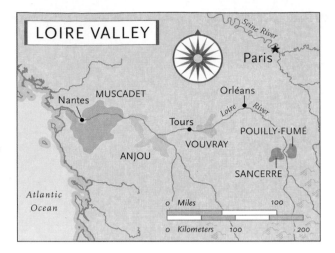

LOIRE VALLEY

Starting at the city of Nantes, a bit upriver from the Atlantic Ocean, the Loire valley stretches inland for six hundred miles along the Loire River.

There are two white grape varieties you should be familiar with:

SAUVIGNON BLANC   CHENIN BLANC

Rather than choosing by grape variety and shipper, as you would in Alsace, choose Loire Valley wines by style and vintage. Here are the main styles:

Pouilly-Fumé: A dry wine that has the most body and concentration of all the Loire Valley wines. It's made with 100 percent Sauvignon Blanc.

Muscadet: A light, dry wine, made from 100 percent Melon de Bourgogne grapes.

Sancerre: Striking a balance between full-bodied Pouilly-Fumé and light-bodied Muscadet, it's made with 100 percent Sauvignon Blanc.

Vouvray: The "chameleon" can be dry, semisweet, or sweet. It's made from 100 percent Chenin Blanc.

How did Pouilly-Fumé get its name, and what does *fumé* mean?

Many people ask me if Pouilly-Fumé is smoked, because they automatically associate the word *fumé* with smoke. One of the many theories about the origin of the word comes from the white morning mist that blankets the area. As the sun burns off the mist, it looks as if smoke is rising.

When are the wines ready to drink?

Generally, Loire Valley wines are meant to be consumed young. The exception is a sweet Vouvray, which can be laid down for a longer time.

Here are more specific guidelines:

Pouilly-Fumé: Three to five years
Sancerre: Two to three years
Muscadet: One to two years

What's the difference between Pouilly-Fumé and Pouilly-Fuissé?

My students often ask me this, expecting similarly named wines to be related. But Pouilly-Fumé is made from 100 percent Sauvignon Blanc and comes from the Loire Valley, while Pouilly-Fuissé is made from 100 percent Chardonnay and comes from the Mâconnais region of Burgundy.

What are the most recent trends in Loire Valley wines?

When I started in the wine business, the most important Loire Valley wine was Pouilly-Fumé. Back then, Sancerre played a minor role and cost much less than Pouilly-Fumé. Over the last twenty years, Sancerre's popularity has grown. Today, Sancerre is sometimes even more expensive than its neighbor. Both wines are made from the same grape variety, 100 percent Sauvignon Blanc, and have a similar style, most being unoaked and medium-bodied with great acidity and fruit balance.

Muscadet, on the other hand, remains a good value. In fact, these wines are being made even better than they were twenty years ago yet their prices remain affordable.

BEST BETS FOR RECENT VINTAGES OF THE LOIRE VALLEY

2002*   2003   2004   2005*   2006

*Note: * signifies exceptional vintage*

KEVIN ZRALY'S FAVORITE PRODUCERS

Sancerre: Archambault, Roblin, Lucien Crochet, Jean Vacheron, Château de Sancerre, Domaine Fournier, Henri Bourgeois

Pouilly-Fumé: Guyot, Michel Redde, Ch. de Tracy, Dagueneau, Ladoucette, Colin, Jolivet, Jean-Paul Balland

Vouvray: Huet

Muscadet: Marquis de Goulaine, Sauvion, Métaireau

Savennières: Nicolas Joly

IF YOU see the phrase "sur lie" on a Muscadet wine label, it means that the wine was aged on its lees (sediment).

THE LOIRE VALLEY also produces the world-famous Anjou Rosé.

WINE AND FOOD

BARON PATRICK LADOUCETTE, *whose winery is the largest producer of Pouilly-Fumé, suggests the following combinations:*

Pouilly-Fumé: *"Smoked salmon; turbot with hollandaise; white meat chicken; veal with cream sauce."*

Sancerre: *"Shellfish, simple food of the sea, because Sancerre is drier than Pouilly-Fumé."*

Muscadet: *"All you have to do is look at the map to see where Muscadet is made: by the sea where the main fare is shellfish, clams, and oysters."*

Vouvray: *"A nice semidry wine to have with fruit and cheese."*

MARQUIS ROBERT DE GOULAINE *suggests these combinations:*

Muscadet: *"Muscadet is good with a huge variety of excellent and fresh 'everyday' foods, including all the seafood from the Atlantic Ocean, the fish from the river—pike, for instance—game, poultry, and cheese (mainly goat cheese). Of course, there is a must in the region of Nantes: freshwater fish with the world-famous butter sauce, the beurre blanc, invented at the turn of the century by Clémence, who happened to be the chef at Goulaine. If you prefer, try Muscadet with a dash of crème de cassis (black currant); it is a wonderful way to welcome friends!"*

GRAND CRU CLASSÉ DE GRAVES

Château Olivier

PESSAC-LÉOGNAN
APPELLATION PESSAC-LÉOGNAN CONTROLÉE

GROUPEMENT FONCIER AGRICOLE DU CHATEAU OLIVIER
J.-J. DE BETHMANN - PROPRIÉTAIRE
LEOGNAN - 33850 - GIRONDE - FRANCE

MIS EN BOUTEILLE AU CHATEAU

12,5% vol   PRODUCE OF FRANCE   75 cl

BORDEAUX PRODUCTION:
   84% red
   16% white

THE NAME "Graves" means
"gravel"—the type of soil
found in the region.

WHEN PEOPLE think of dry
white Bordeaux wines, they
normally think of the major
areas of Graves or Pessac-
Léognan, but some of the
best value/quality white
wines produced in Bordeaux
come from the Entre-Deux-
Mers area.

CLASSIFIED WHITE châteaux
wines make up only 3% of
the total production of white
Graves.

THE WHITE WINES OF BORDEAUX

Doesn't Bordeaux always mean red wine?

That's a misconception. Actually, two of the five major areas of Bordeaux—
Graves and Sauternes—are known for their excellent white wines. Sauternes
is world-famous for its sweet white wine.

   The major white grape varieties used in both areas are:

SAUVIGNON BLANC   **SÉMILLON**

GRAVES

How are the white Graves wines classified?

There are two levels of quality:

GRAVES   **PESSAC-LÉOGNAN**

   The most basic Graves is simply called "Graves." The best wines are produced
in Pessac-Léognan. Those labeled "Graves" are from the southern portion of the

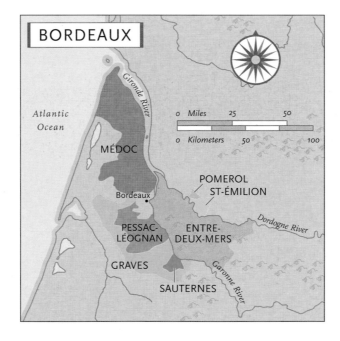

region surrounding Sauternes,
while Pessac-Léognan is in the
northern half of the region, next
to the city of Bordeaux. The best
wines are known by the name of
a particular château, a special
vineyard that produces the best-
quality grapes. The grapes grown
for these wines enjoy better soil
and better growing conditions
overall. The classified château
wines and the regional wines of
Graves are always dry.

How should I select a Graves wine?

My best recommendation would be to purchase a classified château wine. The châteaux are:

CHÂTEAU HAUT-BRION

CHÂTEAU CARBONNIEUX*

CHÂTEAU COUHINS-LURTON

CHÂTEAU LA TOUR-MARTILLAC

CHÂTEAU MALARTIC-LAGRAVIÈRE

CHÂTEAU SMITH-HAUT-LAFITTE

CHÂTEAU BOUSCAUT*

DOMAINE DE CHEVALIER

CHÂTEAU LA LOUVIÈRE*

CHÂTEAU LAVILLE-HAUT-BRION

CHÂTEAU OLIVIER*

The largest producers and the easiest to find.

What are the most recent trends in the white wines of Bordeaux?

In the last twenty years, out of all of the white wine regions of France, the most significant shifts have occurred in Bordeaux. The winemakers have changed the style with more modern winemaking techniques and by being more careful with their selection in the vineyard, thus resulting in much higher quality white Bordeaux wines.

THE STYLE of classified white château wines varies according to the ratio of Sauvignon Blanc and Sémillon used. Château Olivier, for example, is made with 65% Sémillon, and Château Carbonnieux with 65% Sauvignon Blanc.

BEST BETS FOR RECENT VINTAGES OF WHITE GRAVES

2000*   2001   2002   2003   2005*

*Note: * signifies exceptional vintage*

WINE AND FOOD

DENISE LURTON-MOULLE (*Château La Louvière, Château Bonnet*): *With Château La Louvière Blanc: grilled sea bass with a beurre blanc, shad roe, or goat cheese soufflé. With Château Bonnet Blanc: oysters on the half-shell, fresh crab salad, mussels, and clams.*

JEAN-JACQUES DE BETHMANN (*Château Olivier*): *"Oysters, lobster, Rouget du Bassin d'Arcachon."*

ANTONY PERRIN (*Château Carbonnieux*): *"With a young Château Carbonnieux Blanc: chilled lobster consommé, or shellfish, such as oysters, scallops, or grilled shrimp. With an older Carbonnieux: a traditional sauced fish course or a goat cheese."*

SAUTERNES/BARSAC

French Sauternes are always sweet, meaning that not all the grape sugar has turned into alcohol during fermentation. A dry French Sauternes doesn't exist. The Barsac district, adjacent to Sauternes, has the option of using Barsac or Sauternes as its appellation.

What are the two different quality levels in style?

1. Regional ($)
2. Classified château ($$$)/($$$$)

Sauternes is still producing one of the best sweet wines in the world. With the great vintages of 2001, 2002, 2003, and 2005, you'll be able to find excellent regional Sauternes if you buy from the best shippers. These wines represent a good value for your money, considering the labor involved in production, but they won't have the same intensity of flavor as a classified château.

What are the main grape varieties in Sauternes?

SÉMILLON   **SAUVIGNON BLANC**

If the same grapes are used for both the dry Graves and the sweet Sauternes, how do you explain the extreme difference in styles?

First and most important, the best Sauternes is made primarily with the Sémillon grape. Second, to make Sauternes, the winemaker leaves the grapes on the vine longer. He waits for a mold called *Botrytis cinerea* ("noble rot") to form. When "noble rot" forms on the grapes, the water within them evaporates and they shrivel. Sugar becomes concentrated as the grapes "raisinate." Then, during the winemaking process, not all the sugar is allowed to ferment into alcohol: hence, the high residual sugar.

WHEN BUYING regional Sauternes look for these reputable shippers: Baron Philippe de Rothschild and B&G.

THERE ARE more Sémillon grapes planted in Bordeaux than there are Sauvignon Blanc.

OTHER SWEET-wine producers in Bordeaux: Ste-Croix-du-Mont and Loupiac.

SAUTERNES IS expensive to produce because several pickings must be completed before the crop is entirely harvested. The harvest can last into November.

How are Sauternes classified?

FIRST GREAT GROWTH—GRAND PREMIER CRU

Château d'Yquem*

FIRST GROWTH—PREMIERS

Château La Tour Blanche*   Château Lafaurie-Peyraguey*
Clos Haut-Peyraguey*   Château de Rayne-Vigneau*
Château Suduiraut*   Château Coutet* (Barsac)
Château Climens* (Barsac)   Château Guiraud*
Château Rieussec*   Château Rabaud-Promis
Château Sigalas-Rabaud*

SECOND GROWTHS—DEUXIÈMES CRUS

Château Myrat (Barsac)   Château Doisy-Daëne (Barsac)
Château Doisy-Védrines* (Barsac)   Château Doisy-Dubroca (Barsac)
Château d'Arche   Château Filhot*
Château Broustet (Barsac)   Château Nairac* (Barsac)
Château Caillou (Barsac)   Château Suau (Barsac)
Château de Malle*   Château Romer du Hayot*
Château Lamothe   Château Lamothe-Guignard

* These are the châteaux most readily available in the United States.

BEST BETS FOR VINTAGES OF SAUTERNES

1986* 1988* 1989* 1990* 1995 1996 1997* 1998
1999 2000 2001* 2002 2003* 2005* 2006

Note: * signifies exceptional vintage

JUST DESSERTS

My students always ask me, "What do you serve with Sauternes?" Here's a little lesson I learned when I first encountered the wines of Sauternes.

Many years ago, when I was visiting the Sauternes region, I was invited to one of the châteaux for dinner. Upon arrival, my group was offered appetizers of foie gras, and, to my surprise, Sauternes was served with it. All the books I had ever read said you should serve drier wines first and sweeter wines later. But since I was a guest, I thought it best not to question my host's selection.

When we sat down for the first dinner course (fish), we were once again served a Sauternes. This continued through the main course—which happened to be rack of lamb—when another Sauternes was served.

Continued . . .

DOESN'T CHÂTEAU d'Yquem make a dry white wine? Yes, it does, and it's simply called "Y." By law, dry wine made in Sauternes cannot be called Appellation Sauternes. It can only be called Appellation Bordeaux.

NOT CLASSIFIED, but of outstanding quality: Château Fargues, Château Gilette, and Château Raymond Lafon.

CHÂTEAU RIEUSSEC is owned by the same family as Château Lafite-Rothschild.

WORLD RECORD

One bottle of Château d'Yquem 1847 sold for $71,675 at the Zachy's wine auction in 2004.

SAUTERNES IS a wine you can age. In fact, most classified château wines in good vintages can easily age for 10 to 30 years.

I thought for sure our host would serve a great old red Bordeaux with the cheese course, but I was wrong again. With the Roquefort cheese was served a very old Sauternes.

With dessert soon on its way, I got used to the idea of having a dinner with Sauternes, and waited with anticipation for the final choice. You can imagine my surprise when a dry red Bordeaux—Château Lafite-Rothschild—was served with dessert!

Their point was that Sauternes doesn't have to be served only with dessert. All the Sauternes went well with the courses, because all the sauces complemented the wine and food.

By the way, the only wine that didn't go well with dinner was the Château Lafite-Rothschild with dessert, but we drank it anyway!

Perhaps this anecdote will inspire you to serve Sauternes with everything. Personally, I prefer to enjoy Sauternes by itself; I'm not a believer in the "dessert wine" category. This dessert wine is dessert in itself.

THE WHITE WINES OF BURGUNDY

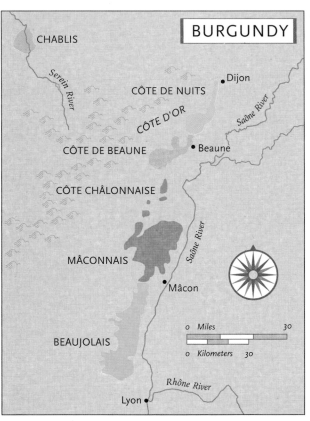

BURGUNDY

CHABLIS

Serein River

CÔTE DE NUITS

Dijon

CÔTE D'OR

Saône River

CÔTE DE BEAUNE

Beaune

CÔTE CHÂLONNAISE

Saône River

MÂCONNAIS

Mâcon

0   Miles   30

0   Kilometers   30

BEAUJOLAIS

Rhône River

Lyon

Where is Burgundy?

Burgundy is a region located in central eastern France. Its true fame is as one of the finest wine-producing areas in the world.

What is Burgundy?

Burgundy is one of the major wine-producing regions that hold an AOC designation in France. However, over the years, I have often found that people are confused about what a Burgundy really is, because the name has been borrowed so freely.

Burgundy is *not* a synonym for red wine, even though the color known as burgundy is obviously named after red wine. Many of the world's most renowned (and expensive) white wines come from Burgundy. Adding to the confusion (especially going back twenty-plus years) is that many red wines around the world were simply labeled "Burgundy" even though they were ordinary table wines. There are still some wineries, especially in the United States, that continue to label their wines as Burgundy, but these wines have no resemblance to the style of authentic French Burgundy wines.

What are the main regions within Burgundy?

CHABLIS   CÔTE D'OR}   CÔTE DE NUITS
                             CÔTE DE BEAUNE

CÔTE CHÂLONNAISE   MÂCONNAIS   BEAUJOLAIS

Before we explore Burgundy region by region, it's important to know the types of wine that are produced there. Take a look at the chart below: It breaks down the types of wine and tells you the percentage of reds to whites.

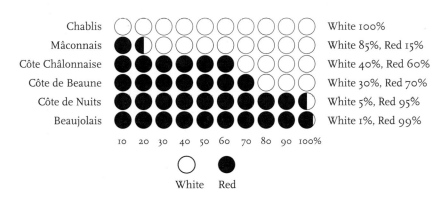

Chablis	White 100%
Mâconnais	White 85%, Red 15%
Côte Châlonnaise	White 40%, Red 60%
Côte de Beaune	White 30%, Red 70%
Côte de Nuits	White 5%, Red 95%
Beaujolais	White 1%, Red 99%

10 20 30 40 50 60 70 80 90 100%

○ White   ● Red

Burgundy is another region so famous for its red wines that people may forget that some of the finest white wines of France are also produced there. The three areas in Burgundy that produce world-famous white wines are:

CHABLIS   CÔTE DE BEAUNE   MÂCONNAIS

If it's any comfort to you, you need to know only one white grape variety: Chardonnay. All the great white Burgundies are made from 100 percent Chardonnay.

Is there only one type of white Burgundy?

Although Chardonnay is used to make all the best French white Burgundy wines, the three areas produce many different styles. Much of this has to do with where the grapes are grown and the vinification procedures. For example, the northern climate of Chablis produces wines with more acidity than those in the southern region of Mâconnais.

THE LARGEST city in Burgundy is known not for its wines but for another world-famous product. The city is Dijon, and the product is mustard.

ALTHOUGH CHABLIS is part of the Burgundy region, it is a three-hour drive south from there to the Mâconnais area.

CÔTE D'OR PRODUCTION:
  78% red
  22% white

ANOTHER WHITE grape found in the Burgundy region is the Aligoté. It is a lesser grape variety and the grape name usually appears on the label.

THE STORY OF KIR

The apéritif Kir has been popular from time to time. It is a mixture of white wine and crème de cassis (made from black currants). It was the favorite drink of the former mayor of Dijon, Canon Kir, who originally mixed in the sweet cassis to balance the high acidity of the local white wine made from the Aligoté grape.

With regard to vinification procedures, after the grapes are harvested in the Chablis and Mâconnais areas, most are fermented and aged in stainless-steel tanks. In the Côte de Beaune, after the grapes are harvested, a good percentage of the wines are fermented in small oak barrels and also aged in oak barrels. The wood adds complexity, depth, body, flavor, and longevity to the wines.

White Burgundies have one trait in common: They are dry.

How are the white wines of Burgundy classified?

The type of soil and the angle and direction of the slope are the primary factors determining quality. Here are the levels of quality:

Village Wine: Bears the name of a specific village. ($ = good)

Premier Cru: From a specific vineyard with special characteristics, within one of the named villages. Usually a Premier Cru wine will list on the label the village first and the vineyard second. ($$ = better)

Grand Cru: From a specific vineyard that possesses the best soil and slope in the area and meets or exceeds all other requirements. In most areas of Burgundy, the village doesn't appear on the label—only the Grand Cru vineyard name is used. ($$$$ = best)

ALSO LOOK for regional Burgundy wine, such as Bourgogne Blanc or Bourgogne Rouge.

MOST PREMIER Cru wines give you the name of the vineyard on the label, but others are simply called "Premier Cru," which is a blend of different cru vineyards.

THE AVERAGE yield for a village wine in Burgundy is 360 gallons per acre. For the Grand Cru wines it is 290 gallons per acre, a noticeably larger concentration, which produces a more flavorful wine.

A NOTE ON THE USE OF WOOD

Each wine region in the world has its own way of producing wines. Wine was always fermented and aged in wood—until the introduction of cement tanks, glass-lined tanks, and, most recently, stainless-steel tanks. Despite these technological improvements, many winemakers prefer to use the more traditional methods. For example, some of the wines from the firm of Louis Jadot are fermented in wood as follows:

One-third of the wine is fermented in new wood.

One-third of the wine is fermented in year-old wood.

One-third of the wine is fermented in older wood.

Jadot's philosophy is that the better the vintage, the newer the wood: Younger wood imparts more flavor and tannin, which might overpower wines of lesser vintage. Thus, younger woods are generally reserved for aging the better vintages.

CHABLIS

Chablis is the northernmost area in Burgundy, and it produces only white wine.

Isn't Chablis just a general term for white wine?

The name "Chablis" suffers from the same misinterpretation and overuse as does the name "Burgundy." Because the French didn't take the necessary precautions to protect the use of the name, "Chablis" is now randomly applied to many ordinary bulk wines from other countries. Chablis has come to be associated with some very undistinguished wines, but this is not the case with French Chablis. In fact, the French take their Chablis very seriously. There are special classification and quality levels for Chablis.

What are the quality levels of Chablis?

Petit Chablis: The most ordinary Chablis; rarely seen in the United States.
Chablis: A wine that comes from grapes grown anywhere in the Chablis district.
Chablis Premier Cru: A very good quality of Chablis that comes from specific high-quality vineyards.
Chablis Grand Cru: The highest classification of Chablis, and the most expensive because of its limited production. There are only seven vineyards in Chablis entitled to be called "Grand Cru."

ALL FRENCH Chablis is made of 100% Chardonnay grapes.

THERE ARE more than 250 grape growers in Chablis, but only a handful age their wine in wood.

OF THESE quality levels, the best value is a Chablis Premier Cru.

THERE ARE only 245 acres planted in Grand Cru vineyards.

VILLAGE $

PREMIER CRU $$

GRAND CRU $$$$

THE WINTER temperatures in some parts of Chablis can match those of Norway.

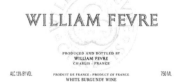

If you're interested in buying only the best Chablis, here are the seven Grands Crus and the most important Premiers Cru vineyards:

<hr>

SOME OF THE GRAND CRU VINEYARDS OF CHABLIS

<hr>

Blanchots	Preuses
Bougros	Valmur
Grenouilles	Vaudésir
Les Clos	

<hr>

SOME OF THE TOP PREMIER CRU VINEYARDS OF CHABLIS

<hr>

Côte de Vaulorent	Montmains
Fourchaume	Monts de Milieu
Lechet	Vaillon
Montée de Tonnerre	

What has been the most important recent change in Chablis?

The cold, northern climate of Chablis poses a threat to the vines. Back in the late 1950s, Chablis almost went out of business because the crops were ruined by frost. Through modern technology, and with improved methods of frost protection, vintners have learned to control this problem, so more and better wine is being produced.

How should I buy Chablis?

The two major aspects to look for in Chablis are the shipper and the vintage. Here is a list of the most important shippers of Chablis to the United States:

A. REGNARD & FILS	JOSEPH DROUHIN
ALBERT PIC & FILS	LA CHABLISIENNE
DOMAINE LAROCHE	LOUIS JADOT
FRANÇOIS RAVENEAU	LOUIS MICHEL
GUY ROBIN	RENÉ DAUVISSAT
J. MOREAU & FILS	ROBERT VOCORET
JEAN DAUVISSAT	WILLIAM FÈVRE

BEST BETS FOR RECENT VINTAGES OF CHABLIS

1999   2000*   2001   2002*   2004   2005*

*Note: * signifies exceptional vintage*

When should I drink my Chablis?

Chablis: Within two years of the vintage.
Premier Cru: Between two and four years of the vintage.
Grand Cru: Between three and eight years of the vintage.

CÔTE DE BEAUNE

This is one of the two major areas of the Côte d'Or. The wines produced here are some of the finest examples of dry white Chardonnay produced in the world and are considered a benchmark for winemakers everywhere.

THE LARGEST Grand Cru, in terms of production, is Corton-Charlemagne, which represents more than 50% of all white Grand Cru wines.

CÔTE DE BEAUNE

Here is a list of my favorite white wine–producing villages and vineyards in the Côte de Beaune.

VILLAGE	PREMIER CRU VINEYARDS	GRAND CRU VINEYARDS
Aloxe-Corton		Corton-Charlemagne
		Charlemagne
Beaune	Clos des Mouches	None
Meursault	Les Perrières	None
	Les Genevrières	
	La Goutte d'Or	
	Les Charmes	
	Blagny	
	Poruzots	
Puligny-Montrachet	Les Combettes	Montrachet*
	Les Caillerets	Bâtard-Montrachet*
	Les Pucelles	Chevalier-Montrachet
	Les Folatières	Bienvenue-Bâtard-Montrachet
	Clavoillons	
	Les Referts	
Chassagne-Montrachet	Les Ruchottes	Montrachet*
	Morgeot	Bâtard-Montrachet*
		Criots-Bâtard-Montrachet

* The vineyards of Montrachet and Bâtard-Montrachet overlap between the villages of Puligny-Montrachet and Chassagne-Montrachet.

VILLAGE $	PREMIER CRU $$	GRAND CRU $$$
Puligny-Montrachet APPELLATION CONTROLÉE — *Louis Latour* MIS EN BOUTEILLE PAR LOUIS LATOUR NÉGOCIANT A BEAUNE (COTE-D'OR)	*Puligny-Montrachet* LES REFERTS APPELLATION CONTROLÉE — *Louis Latour* MIS EN BOUTEILLE PAR LOUIS LATOUR NÉGOCIANT A BEAUNE (COTE-D'OR), FRANCE	*Montrachet* APPELLATION CONTROLÉE LOUIS LATOUR, NÉGOCIANT A BEAUNE (CÔTE-D'OR), FRANCE

The three most important white wine–producing villages of the Côte de Beaune are Meursault, Puligny-Montrachet, and Chassagne-Montrachet. All three produce their white wine from the same grape—100 percent Chardonnay.

"The difference between the Village wine, Puligny-Montrachet, and the Grand Cru Montrachet is not in the type of wood used in aging or how long the wine is aged in wood. The primary difference is in the location of the vineyards, i.e., the soil and the slope of the land."

—ROBERT DROUHIN

What makes each Burgundy wine different?

In Burgundy, one of the most important factors in making a good wine is soil. The quality of the soil is the main reason why there are three levels and price points between a Village, a Premier Cru, and a Grand Cru wine. Another major factor that differentiates each wine is the vinification procedure the winemaker uses—the recipe. It's the same as if you were to compare the chefs at three gourmet restaurants: They may start out with the same ingredients, but it's what they do with those ingredients that matters.

BEST BETS FOR CÔTE DE BEAUNE WHITE

1995* 1996* 1999 2000* 2001 2002* 2004* 2005*

Note: * signifies exceptional vintage

CÔTE CHÂLONNAISE

The Côte Châlonnaise is the least known of the major wine districts of Burgundy. Although the Châlonnaise produces such red wines as Givry and Mercurey (see Class Four, "The Red Wines of Burgundy and the Rhône Valley"), it also produces some very good white wines that not many people are familiar

GRAND VIN DE BOURGOGNE
Produit de France — Produce of France
Rully 1er Cru
APPELLATION RULLY 1ᵉʳ CRU CONTRÔLÉE
VINIFIÉ ÉLEVÉ ET MIS EN BOUTEILLES PAR
Olivier Leflaive
ALC 13% BY VOLUME   SEVIB S.A. 21190 PULIGNY-MONTRACHET   L - RU1 961   CONTENTS 750 ml

with, which means value for you. I'm referring to the wines of Montagny and Rully. These wines are of the highest quality produced in the area, similar to the white wines of the Côte d'Or but less costly.

Look for the wines of Antonin Rodet, Faiveley, Louis Latour, Moillard, and Olivier Leflaive.

MÂCONNAIS

The southernmost white wine–producing area in Burgundy, the Mâconnais has a climate warmer than that of the Côte d'Or and Chablis. Mâcon wines are, in general, pleasant, light, uncomplicated, reliable, and a great value.

What are the quality levels of Mâconnais wines?

From basic to best:

1. **MÂCON BLANC**

2. **MÂCON SUPÉRIEUR**

3. **MÂCON-VILLAGES**

4. **ST-VÉRAN**

5. **POUILLY-VINZELLES**

6. **POUILLY-FUISSÉ**

Of all Mâcon wines, Pouilly-Fuissé is unquestionably one of the most popular. It is among the highest-quality Mâconnais wines, fashionable to drink in the United States long before most Americans discovered the splendors of wine. As wine consumption increased in America, Pouilly-Fuissé and other famous areas such as Pommard, Nuits-St-Georges, and Chablis became synonymous with the best wines of France and could always be found on any restaurant's wine list.

In my opinion, Mâcon-Villages is the best value. Why pay more for Pouilly-Fuissé—sometimes three times as much—when a simple Mâcon will do just as nicely?

BEST BETS FOR RECENT VINTAGES OF MÂCON WHITE

2002* 2003 2004 2005*

*Note: * signifies exceptional vintage*

MORE THAN four-fifths of the wines from the Mâconnais are white.

THERE IS a village named Chardonnay in the Mâconnais area, where the grape's name is said to have originated.

Joseph Drouhin

POUILLY-FUISSÉ

APPELLATION CONTROLÉE
MIS EN BOUTEILLE PAR JOSEPH DROUHIN NÉGOCIANT ÉLEVEUR À BEAUNE, CÔTE - D'OR, FRANCE, AUX CELLIERS DES ROIS DE FRANCE ET DES DUCS DE BOURGOGNE ALC 13.0% BY VOL

SOLE AGENTS Dreyfus, Ashby & Co NEW YORK, N.Y.

IN AN average year, around 450,000 cases of Pouilly-Fuissé are produced—not nearly enough to supply all the restaurants and retail shops for worldwide consumption.

SINCE MÂCON wines are usually not aged in oak, they are ready to drink as soon as they are released.

If you're taking a client out on a limited expense account, a safe wine to order is Mâcon. If the sky's the limit, go for the Meursault!

OVERVIEW

Now that you're familiar with the many different white wines of Burgundy:

How do you choose the right one for you?

First look for the vintage year. With Burgundy, it's especially important to buy a good year. After that, your choice becomes a matter of taste and cost. If price is no object, aren't you lucky?

Also, after some trial and error, you may find that you prefer the wines of one shipper over another. Here are some of the shippers to look for when buying white Burgundy:

BOUCHARD PÈRE & FILS
CHANSON
JOSEPH DROUHIN
LABOURÉ-ROI
LOUIS JADOT
LOUIS LATOUR
MOMMESSIN
OLIVIER LEFLAIVE FRÈRES
PROSPER MAUFOUX
ROPITEAU FRÈRES

Although 80 percent of Burgundy wines are sold through shippers, some fine estate-bottled wines are available in limited quantities in the United States. The better ones include:

CHÂTEAU FUISSÉ (POUILLY-FUISSÉ)
DOMAINE BACHELET-RAMONET (CHASSAGNE-MONTRACHET)
DOMAINE BOILLOT (MEURSAULT)
DOMAINE BONNEAU DU MARTRAY (CORTON-CHARLEMAGNE)
DOMAINE COCHE-DURY (MEURSAULT, PULIGNY-MONTRACHET)
DOMAINE DES COMTES LAFON (MEURSAULT)
DOMAINE ÉTIENNE SAUZET (CHASSAGNE-MONTRACHET,
    PULIGNY-MONTRACHET)
DOMAINE LEFLAIVE (MEURSAULT, PULIGNY-MONTRACHET)
DOMAINE MATROT (MEURSAULT)
DOMAINE VINCENT GIRARDIN (CHASSAGNE-MONTRACHET,
    PULIGNY-MONTRACHET)

DOMAINE LEFLAIVE's wines are named for characters and places in a local medieval tale. The Chevalier of Puligny-Montrachet, lonely for his son who was off fighting in the Crusades, amused himself in the ravine-like vineyards (Les Combettes) with a local maiden (Pucelle), only to welcome the arrival of another son (Bâtard-Montrachet) nine months later.

ESTATE-BOTTLED wine: The wine is made, produced, and bottled by the owner of the vineyard.

What are the recent trends in white Burgundy?

If anything, white Burgundy wines have gotten better over the years. Since 1995—and especially with the vintages 2002 and 2005—Burgundy has made some of its best wines ever. Mâcon wines remain one of today's great values, being made from 100 percent Chardonnay grapes, yet usually priced to under fifteen dollars a bottle.

One of the most interesting things I have seen on the labels of French wines in the U.S. market, particularly from the Mâconnais, is the inclusion of the grape variety. French winemakers have finally realized that Americans buy wines by grape varieties.

I have also found that the major shippers continue to make high-quality wines.

FOR FURTHER READING

I recommend *Burgundy* by Anthony Hanson; *Burgundy* by Robert M. Parker Jr.; *The Great Domaines of Burgundy* by Remington Norman; and *Making Sense of Burgundy* by Matt Kramer.

WINE AND FOOD

When you choose a white Burgundy wine, you have a whole gamut of wonderful food possibilities. Let's say that you decide upon a wine from the Mâconnais area: Very reasonably priced, Mâconnais wines are suitable for picnics, as well as for more formal dinners. Or, you might select one of the fuller-bodied Côte de Beaune wines, or, if you prefer, an all-purpose Chablis that can even stand up to a hearty steak. Here are some tempting combinations offered by the winemakers.

ROBERT DROUHIN: *With a young Chablis or St-Véran, Mr. Drouhin enjoys shellfish. "Fine Côte d'Or wines match well with any fish or light white meat such as veal or sweetbreads. But please, no red meat."*

CHRISTIAN MOREAU: *"A basic Village Chablis is good as an apéritif and with hors d'oeuvres and salads. A great Premier Cru or Grand Cru Chablis needs something more special, such as lobster. It's an especially beautiful match if the wine has been aged a few years."*

PIERRE HENRY GAGEY *(Louis Jadot): "My favorite food combination with white Burgundy wine is, without doubt, homard grillé Breton (blue lobster). Only harmonious, powerful, and delicate wines are able to go with the subtle, thin flesh and very fine taste of the Breton lobster."*

Mr. Gagey says Chablis is a great match for oysters, snails, and shellfish, but a "Grand Cru Chablis should be had with trout."

On white wines of the Côte de Beaune,

Mr. Gagey gets a bit more specific: "With Village wines, which should be had at the beginning of the meal, try a light fish or quenelles (light dumplings). Premier Cru and Grand Cru wines can stand up to heavier fish and shellfish such as lobster—but with a wine such as Corton-Charlemagne, smoked Scottish salmon is a tasty choice."

Mr. Gagey's parting words on the subject: "Never with meat."

LOUIS LATOUR: *"With Corton-Charlemagne, filet of sole in a light Florentine sauce. Otherwise, the Chardonnays of Burgundy complement roast chicken, seafood, and light-flavored goat cheese particularly well." Mr. Latour believes that one should have Chablis with oysters and fish.*

Questions for Class One:
The White Wines of France

1. Match grape variety with wine region.

    a. Riesling   ____Champagne

    b. Sauvignon Blanc   ____Loire

    c. Chardonnay   ____Alsace

    d. Semillon   ____Burgundy

    e. Gewurtraminer   ____Bordeaux

    f. Grenache   ____Côte du Rhone

    g. Pinot Noir

    h. Cabernet Sauvignon

    i. Chenin Blanc

    j. Syrah

    k. Merlot

2. When were the Appellation d'Origine Controlée (AOC) laws established?

3. How many acres are there in a hectare?

4. How many gallons are there in a hectoliter?

5. What is one difference in style between a Riesling from Alsace and a Riesling from Germany?

6. What is the most planted grape in Alsace?

7. Name two important shippers of Alsace wine.

8. What is the grape variety for the wines Sancerre and Pouilly Fumé?

9. What is the grape variety for the wine Vouvray?

The Wines of Washington, Oregon, and New York

The White Wines of California

WASHINGTON STATE • OREGON • NEW YORK • NATIVE AMERICAN, EUROPEAN,

AND FRENCH-AMERICAN VARIETIES • INTRODUCTION TO CALIFORNIA WINES •

THE WHITE WINES OF CALIFORNIA

The "Big Four" of U.S. Winemaking

THE PACIFIC NORTHWEST winegrowing region includes Washington, Oregon, British Columbia, and Idaho.

WASHINGTON STATE has more Riesling planted than any other state.

Chateau Ste. Michelle is the world's largest Riesling producer.

Now that we have looked at the history and "big picture" of American winemaking, we can turn our attention to the wines themselves. Before we tackle the exciting world of California wines, I'd like to talk about three other major winemaking regions in the United States: Washington State, Oregon, and New York State.

WASHINGTON STATE

In Washington, the climatic conditions are a little cooler and rainier than in California, but it's neither too cold nor too wet to make great wine. The winegrowing regions are protected from Washington's infamous rains by the Cascade Mountains. The earliest record of grape growing in Washington can be traced back to 1825, while the beginning of its modern winemaking industry can be dated to 1967, with the first wine produced under the Chateau Ste. Michelle label.

The four major white grapes grown in Washington are Chardonnay, Riesling, Gewürztraminer, and Sauvignon Blanc. In the 1960s and 1970s, Washington was known only for white wines, but now Washington State has become known as one of the great American states for the production of red wines, especially Merlot and Cabernet Sauvignon. (As it turns out, the state's Columbia Valley is on the same latitude as Bordeaux, France.) Over the last ten years Washington winemakers have increased their plantings of Syrah, and the recent vintages are turning out to be of very high quality.

There are nine AVAs: Columbia Valley, Yakima, Walla Walla, Puget Sound, Red Mountain, Columbia Gorge, Horse Heaven Hills, Wahluke Slope, and Rattlesnake Hills.

AVA (DATE AVA ESTABLISHED)	NUMBER OF WINERIES
Yakima (1983)	50+
Walla Walla (1984)	100+
Columbia Valley* (1984)	100+
Puget Sound (1995)	65+
Red Mountain (2001)	5
Columbia Gorge (2004)	15
Horse Heaven Hills (2005)	4
Wahluke Slope (2006)	3
Rattlesnake Hill (2006)	17

Largest viticultural area, responsible for 95 percent of wine production.

Wineries to look for include Betz, Columbia Crest, Columbia Winery, Hogue Cellars, L'Ecole No. 41, Leonetti Cellars, Woodward Canyon Winery, Andrew Will, Canoe Ridge, Seven Hills, Cayuse, McCrea Cellars, Pepper Bridge, Quilceda Creek, DiStefano, and the largest and oldest winery, Chateau Ste. Michelle.

BEST BETS FOR WASHINGTON WINES

2001*   2002*   2003   2004*   2005*   2006*   2007

*Note: * signifies exceptional vintage*

OREGON

Although grapes were planted and wine was made as early as 1847 in Oregon, the modern era began in the mid-1960s. Today, because of its climate, Oregon is becoming well known for Burgundian-style wines. By Burgundian style I'm referring to Chardonnay and Pinot Noir, the major grapes planted in Oregon. Many critics feel the best Pinot Noir grown in the U.S. is from Oregon. Another success in Oregon is Pinot Gris (aka Pinot Grigio), which has recently overtaken Chardonnay as the most widely planted white grape in the state.

There are nineteen AVAs in Oregon. The major AVA in Oregon is the Willamette Valley, near Portland. Other AVAs include Rogue Valley, Umpqua, and Applegate Valley. Two others, the Columbia Valley and Walla Walla, are AVAs of both Oregon and Washington State.

Wineries to look for include Archery Summit, Argyle, Adelsheim, Beaux Freres, Bergstrom, Bethel Heights, Criston, Domaine Serene, Erath, Eyrie Vineyards, Ken Wright, King Estate, Criston, Ponzi Vineyards, Rex Hill, Shea,

IN 1990 IN Washington, there were fewer than 70 wineries, which turned out less than 2 million cases per year. Today, there are more than 400 wineries, which produce more than 7 million cases. Acreage has tripled.

THE 2005 VINTAGE was the largest harvest ever in Washington State.

WASHINGTON STATE'S wine production is 60% red versus 40% white.

For further reading: *Washington Wines & Wineries, The Essential Guide* by Paul Gregutt.

California	Washington
470,000 acres	31,000+ acres
1,905 wineries	430 wineries
Oregon	**New York**
11,100 acres	30,000 acres
290 wineries	218 wineries

CHATEAU STE. MICHELLE has formed a winemaking partnership with the famous German wine producer Dr. Loosen. The new Riesling wine is called Eroica.

THE LARGEST producer of Pinot Gris (Grigio) in the United States is King Estate in Oregon.

ABOUT 70% OF Oregon's wineries are located in the Willamette Valley.

One of the great wine festivals in the United States is the International Pinot Noir Celebration, which was started in Oregon in 1987.

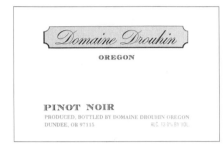

THE 2002 and 2004 VINTAGES are Oregon's best ever.

NEW YORK'S Hudson Valley is one of America's oldest wine-growing regions. French Huguenots planted the grapevines in the 1600s. The Hudson Valley also boasts the oldest active winery in the United States—Brotherhood, which recorded its first vintage in 1839.

THE FIRST winery on Long Island was started in 1973 by Alex and Louisa Hargrave.

THE CLIMATE on Long Island has more than 200 days of sunshine and a longer growing season, making it perfect for Merlot and Bordeaux style wines.

THE THREE AVAs on Long Island are the North Fork, the Hamptons, and Long Island.

THERE ARE more than 200 wineries in New York State, up from just 19 in 1975. In the three major regions, the Finger Lakes region has 73, the Hudson Valley has 28, and Long Island has 33.

Sokol Blosser, Soter, St. Innocent, and Tualatin. Also, the famous Burgundy producer Joseph Drouhin owns a winery in Oregon called Domain Drouhin, producing, not surprisingly, Burgundy-style wines.

BEST BETS FOR OREGON WINES

1999* 2001 2002** 2003 2004* 2005* 2006

Note: * signifies exceptional vintage   ** one of the best vintages ever for Pinot Noir in Oregon

FOR FURTHER READING

I recommend *The Oxford Companion to the Wines of North America* by Bruce Cass and Jancis Robinson; *The Wines of the Pacific Northwest* by Lisa Shara Hall; and *The Northwest Wine Guide* by Andy Perdue.

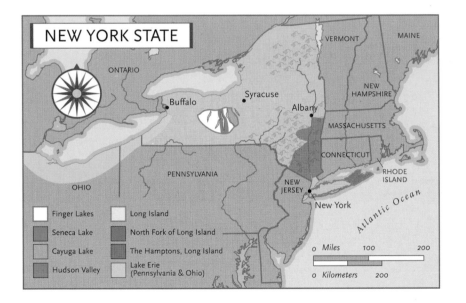

NEW YORK STATE

New York is the third largest wine-producing state in the United States, with nine AVAs. The three premium wine regions in New York are:

Finger Lakes: With the largest wine production east of California.
Hudson Valley: Concentrating on premium farm wineries.
Long Island: New York's red wine region.

Which grapes grow in New York State?

There are three main categories:

> **Native American** (*Vitis labrusca*)
> **European** (*Vitis vinifera*)
> **French-American** (hybrids)

TODAY, MORE than 80 of the 218 New York State wineries produce *vinifera* wines.

THE CONCORD GRAPE represents about two-thirds of New York State's grape acreage and production.

NATIVE AMERICAN VARIETIES

The *Vitis labrusca* vines are very popular among grape growers in New York because they are hardy grapes that can withstand cold winters. Among the most familiar grapes of the *Vitis labrusca* family are Concord, Catawba, and Delaware. Until the last decade, these were the grapes that were used to make most New York wines. In describing these wines, words such as "foxy," "grapey," "Welch's," and "Manischewitz" are often used. These words are a sure sign of *Vitis labrusca*.

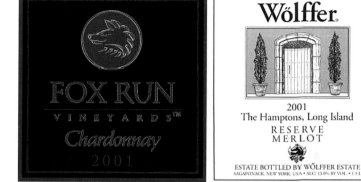

EUROPEAN VARIETIES

Forty years ago, some New York wineries began to experiment with the traditional European (*Vitis vinifera*) grapes. Dr. Konstantin Frank, a Russian viticulturist skilled in cold-climate grape growing, came to the United States and catalyzed efforts to grow *Vitis vinifera* in New York. This was unheard of—and laughed at—back then. Other vintners predicted that he'd fail, that it was impossible to grow *vinifera* in New York's cold and capricious climate.

"What do you mean?" Dr. Frank replied. "I'm from Russia—it's even colder there."

Most people were still skeptical, but Charles Fournier of Gold Seal Vineyards was intrigued enough to give Konstantin Frank a chance to prove his theory. Sure enough, Dr. Frank was successful with the *vinifera* and has produced some world-class wines, especially his Riesling and Chardonnay. So have many other New York wineries, thanks to the vision and courage of Dr. Frank and Charles Fournier.

EXAMPLES OF *Vitis vinifera* grapes are Pinot Grigio, Riesling, Sauvignon Blanc, Chardonnay, Pinot Noir, Merlot, Cabernet Sauvignon, and Syrah.

THE FINGER LAKES wineries produce 85% of New York's wine.

2007 IS the best vintage in New York State since 1995.

FRENCH-AMERICAN VARIETIES

Some New York and East Coast winemakers have planted French-American hybrid varieties, which combine European taste characteristics with American vine hardiness to withstand the cold winters in the Northeast. French viticulturists originally developed these varieties in the nineteenth century. Seyval Blanc and Vidal are the most prominent white wine varieties; Baco Noir and Chancellor are the most common reds.

What are the trends in New York wines?

The most significant developments are taking place on Long Island and in the Finger Lakes region, which have experienced the fastest growth of new vineyards in the state. In the last twenty years, its grape-growing acreage has increased from one hundred acres to more than three thousand acres, with more expansion expected in the future.

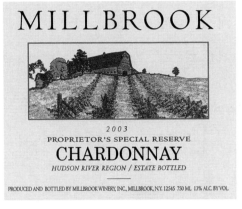

The predominant use of *Vitis vinifera* varieties allows Long Island wineries to compete more effectively in the world market, and Long Island's longer growing season offers greater potential for red grapes.

The Millbrook Winery in the Hudson Valley has shown that this region can produce world-class wines—not only white, but red too, from such grapes as Pinot Noir and Cabernet Franc.

The wines of the Finger Lakes region continue to get better as the winemakers work with grapes that thrive in the cooler climate, including European varieties such as Riesling, Chardonnay, and Pinot Noir.

CALIFORNIA

No winegrowing area in the world has come so far so quickly as California. It seems ironic, because Americans historically have not been very interested in wine. But from the moment Americans first became "wine conscious," winemakers in California rose to the challenge. Thirty years ago we were asking if California wines were entitled to be compared to European wines. Now California wines are available worldwide—exports have increased dramatically in recent years to countries such as Japan, Germany, and England. California produces more than 90 percent of U.S. wine. If the state were a nation, it would be the third leading wine producer in the world!

AN INTRODUCTION TO CALIFORNIA WINES

What are the main viticultural areas of California?

The map on this page should help familiarize you with the wine-making regions. It's easier to remember them if you divide them into four groups:

North Coast: Napa County, Sonoma County, Mendocino County, Lake County (*Best wines: Cabernet Sauvignon, Zinfandel, Sauvignon Blanc, Chardonnay, Merlot*)
North Central Coast: Monterey County, Santa Clara County, Livermore County (*Best wines: Syrah, Grenache, Viognier, Marsanne, Roussane*)
South Central Coast: San Luis Obispo County, Santa Barbara County (*Best wines: Sauvignon Blanc, Chardonnay, Pinot Noir, Syrah*)
San Joaquin Valley (*Known for jug wines*) See page 54.

Although you may be most familiar with the names Napa and Sonoma, less than 10 percent of all California wine comes from these two regions combined. Even so, Napa alone accounts for over 30 percent of dollar sales of California wines. In fact, the bulk of

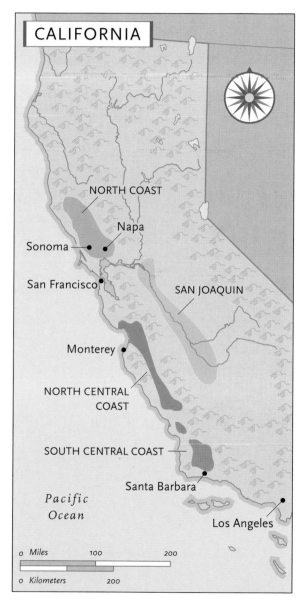

CALIFORNIA WINES dominate American wine consumption, equaling 65% of all sales in the United States.

THERE ARE more than 60,000 wine labels registered in California.

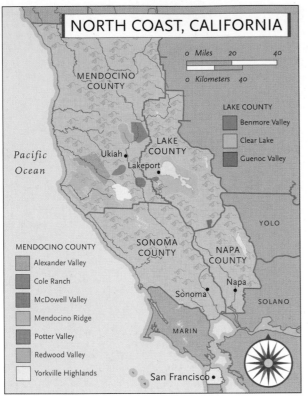

NORTH COAST, CALIFORNIA

MENDOCINO
COUNTY

*Pacific
Ocean*

LAKE
COUNTY

Ukiah •

Lakeport •

LAKE COUNTY

■ Benmore Valley

□ Clear Lake

■ Guenoc Valley

YOLO

SONOMA
COUNTY

NAPA
COUNTY

MENDOCINO COUNTY

■ Alexander Valley

■ Cole Ranch

■ McDowell Valley

■ Mendocino Ridge

■ Potter Valley

■ Redwood Valley

□ Yorkville Highlands

Sonoma •

Napa •

SOLANO

MARIN

San Francisco •

California wine is from the San Joaquin Valley, where mostly jug wines are produced. This region accounts for 58 percent of the wine grapes planted. Maybe that doesn't seem too exciting—that the production of jug wine dominates California winemaking history—but Americans are not atypical in their preferences for this type of wine. In France, for example, AOC wines account for only 35 percent of all French wines, while the rest are everyday table wines.

ACRES OF wine grapes planted in Napa:
44,000

NUMBER OF wineries in Napa: 391

THERE ARE more than 470,000 acres of
vineyards in California.

TOP GRAPES PLANTED IN NAPA

1. Cabernet Sauvignon
2. Merlot
3. Chardonnay

1838: First wine grapes planted in Napa.

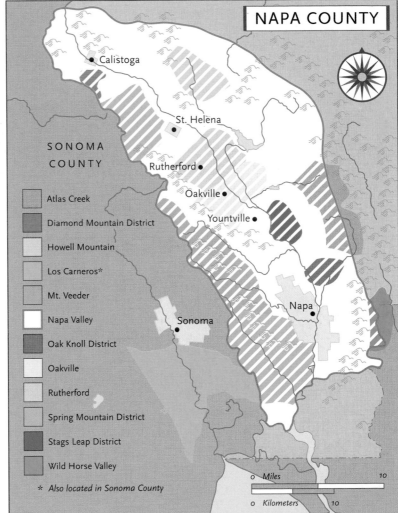

NAPA COUNTY

Calistoga •

St. Helena •

Rutherford •

Oakville •

Yountville •

Napa •

**SONOMA
COUNTY**

□ Atlas Creek

■ Diamond Mountain District

□ Howell Mountain

□ Los Carneros*

□ Mt. Veeder

□ Napa Valley

■ Oak Knoll District

□ Oakville

□ Rutherford

□ Spring Mountain District

■ Stags Leap District

■ Wild Horse Valley

* *Also located in Sonoma County*

Sonoma •

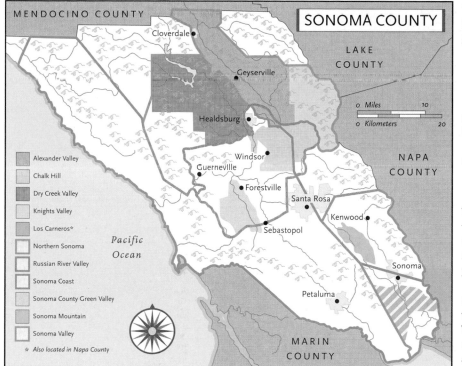

SONOMA COUNTY

MENDOCINO COUNTY

Cloverdale

Geyserville

LAKE
COUNTY

Healdsburg

Windsor

Guerneville

NAPA
COUNTY

Forestville

Santa Rosa

Kenwood

Sebastopol

Sonoma

*Pacific
Ocean*

Petaluma

MARIN
COUNTY

0 Miles 10
0 Kilometers 20

Alexander Valley
Chalk Hill
Dry Creek Valley
Knights Valley
Los Carneros*
Northern Sonoma
Russian River Valley
Sonoma Coast
Sonoma County Green Valley
Sonoma Mountain
Sonoma Valley
* Also located in Napa County

"There is more potential for style variation in California than Europe because of the greater generosity of the fruit."
—WARREN WINIARSKI, *founder, Stag's Leap Wine Cellars, Napa Valley*

ACRES OF wine grapes planted in Sonoma: 60,000

NUMBER OF wineries in Sonoma: 260

TOP GRAPES PLANTED IN SONOMA

1. Chardonnay
2. Pinot Noir
3. Cabernet Sauvignon

LAST YEAR, more than 20 million people visited California wine growing areas. Vineyards and wineries are the second-most popular California tourist destinations after Disneyland!

When did California begin to make better-quality wines?

As early as the 1940s, Frank Schoonmaker, an importer and writer and one of the first American wine experts, convinced some California winery owners to market their best wines using varietal labels.

Robert Mondavi may be one of the best examples of a winemaker who concentrated solely on varietal wine production. In 1966, Mondavi left his family's Charles Krug Winery and started the Robert Mondavi Winery. His role was important to the evolution of varietal labeling of California wines. He was among the first major winemakers to make the total switch that led to higher-quality winemaking.

BIGGER THAN HOLLYWOOD

Wine is California's most valuable finished agricultural product, with a $52 billion economic impact. The film industry: $30 billion.

ROBERT MONDAVI was a great promoter for the California wine industry. "He was able to prove to the public what the people within the industry already knew—that California could produce world-class wines," said Eric Wente.

The majority of California wines retail from $8 to $14.

EARLY CALIFORNIA winemakers sent their children to study oenology at Geisenheim (Germany) or Bordeaux (France). Today, many European winemakers send their children to the University of California at Davis and to Fresno State University.

THE UNIVERSITY of California at Davis graduated only five students from its oenology department in 1966. Today it has a waiting list of students from all over the world.

Examples of new winery technology include:
- heat exchanges
- reverse osmosis
- bladder presses

Clonal selection, wind machines, drip irrigation, biodynamic vineyards, and mechanical harvesting are some new techniques in vineyard management.

A NOTE ON JUG WINES

The phrase jug wine *refers to simple, uncomplicated, everyday drinking wine. You're probably familiar with these types of wine: They're sometimes labeled with a generic name, such as Chablis or Burgundy. Inexpensive and well made, these wines were originally bottled in jugs, rather than in conventional wine bottles, hence the name "jug wine." They are very popular and account for the largest volume of California wine sold in the United States.*

Ernest and Julio Gallo, who began their winery in 1933, are the major producers of jug wines in California. In fact, many people credit the Gallo brothers with converting American drinking habits from spirits to wine. Several other wineries also produce jug wines, among them Almaden, Paul Masson, and Taylor California Cellars.

In my opinion, the best-made jug wines in the world are from California. They maintain both consistency and quality from year to year.

How did California become a world-class producer in just forty years?

There are many reasons for California's winemaking success, including:

Location—Napa and Sonoma counties, two of the major quality-wine regions, are both less than a two-hour drive from San Francisco. The proximity of these regions to the city encourages both residents and tourists to visit the wineries in the two counties, most of which offer wine tastings and sell their wines in their own shops.

Weather—Abundant sunshine, warm daytime temperatures, cool evenings, and a long growing season all add up to good conditions for growing many grape varieties. California is certainly subject to sudden changes in weather, but a fickle climate is not a major worry.

The University of California at Davis and Fresno State University—Both schools have been the training grounds for many young California winemakers, and their curricula concentrate on the scientific study of wine, viticulture, and, most important, technology. Their research, focused on soil, different strains of yeast, hybridization, temperature-controlled fermentation, and other viticultural techniques, has revolutionized the wine industry worldwide.

EARLY PIONEERS

Some of the pioneers of the back-to-the-land movement:

"FARMER"	WINERY	ORIGINAL PROFESSION
Robert Travers	Mayacamas	Investment banker
David Stare	Dry Creek	Civil engineer
Tom Jordan	Jordan	Geologist
James Barrett	Chateau Montelena	Attorney
Tom Burgess	Burgess	Air Force pilot
Jess Jackson	Kendall-Jackson	Attorney
Warren Winiarski	Stag's Leap	College professor
Brooks Firestone	Firestone	Take a guess!

Money and Marketing Strategy—This cannot be overemphasized. Marketing may not make the wine, but it certainly helps sell it. As more and more winemakers concentrated on making the best wine they could, American consumers responded with appreciation. They were willing to buy—and pay—more as quality improved. In order to keep up with consumer expectations, winemakers realized that they needed more research, development, and—most important—working capital. The wine industry turned to investors, both corporate and individual.

Since 1967, when the now-defunct National Distillers bought Almaden, multinational corporations have recognized the profit potential of large-scale winemaking and have aggressively entered the wine business. They've brought huge financial resources and expertise in advertising and promotion that have helped promote American wines domestically and internationally. Other early corporate participants included Pillsbury and Coca-Cola.

On the other side of the investor scale are the individual investor/growers drawn to the business by their love of wine and their desire to live the winemaking "lifestyle." These individuals are more focused on producing quality wines.

Both corporate and individual investors had, by the 1990s, helped California fine-tune its wine industry, which today produces not only delicious and reliable wines in great quantity but also truly outstanding wines, many with investment potential.

SO YOU WANT to buy a vineyard in California? Today, one acre in the Napa Valley costs between $150,000 to $250,000 unplanted, and it takes an additional $20,000 per acre to plant. This per-acre investment sees no return for three to five years. To this, add the cost of building the winery, buying the equipment, and hiring the winemaker.

RECORD PRICE

In 2002, Francis Ford Coppola, owner of Niebaum-Coppola Wine Estate, paid $350,000 an acre for vineyard land in Napa.

IN 1970 the average price per acre in Napa was $5,000.

WHAT ARE the advantages of being a California winery owner?
1. You can wear all the latest styles from the L.L. Bean catalog.
2. You can bone up a lot on ecology and biodynamic farming.
3. You can grow a beard, drive a pickup truck, and wear suspenders—and drive your Bentley on the weekends.

CREATIVE FINANCING

Overheard at a restaurant in Yountville, Napa: "How do you make a small fortune in the wine business?" "Start with a large fortune and buy a winery."

IN CALIFORNIA many winemakers move around from one winery to another, just as good chefs move from one restaurant to the next. This is not uncommon. They may choose to carry and use the same "recipe" from place to place, if it is particularly successful, and sometimes they will experiment and create new styles.

I'VE MENTIONED stainless-steel fermentation tanks so often, I'll give a description, in case you need one. These tanks are temperature controlled, allowing winemakers to control the temperature at which the wine ferments. For example, a winemaker could ferment wines at a low temperature to retain fruitiness and delicacy, while preventing browning and oxidation.

AMBASSADOR ZELLERBACH, who created Hanzell Winery, was one of the first California winemakers to use small French oak aging barrels because he wanted to re-create a Burgundian style.

HOLLYWOOD AND VINE

What do comedians the Smothers Brothers, actors Christina Crawford, Fess Parker, Wayne Rogers, and director Francis Ford Coppola have in common? They all own vineyards in California.

What's meant by *style*? How are different styles of California wine actually created?

Style refers to the characteristics of the grapes and wine. It is the trademark of the individual winemaker—an "artist" who tries different techniques to explore the fullest potential of the grapes.

Most winemakers will tell you that 95 percent of winemaking is in the quality of the grapes they begin with. The other 5 percent can be traced to the "personal touch" of the winemaker. Here are just a few of the hundreds of decisions a winemaker must make when developing his or her style of wine:

- When should the grapes be harvested?

- Should the juice be fermented in stainless-steel tanks or oak barrels? How long should it be fermented? At what temperature?

- Should the wine be aged at all? How long? If so, should it be aged in oak? What kind of oak—American, French?

- What varieties of grape should be blended, and in what proportion?

- How long should the wine be aged in the bottle before it is sold?

The list goes on. Because there are so many variables in winemaking, producers can create many styles of wine from the same grape variety—so you can choose the style that suits your taste. With the relative freedom of winemaking in the United States, the "style" of California wines continues to be "diversity."

Why is California wine so confusing?

The renaissance of the California wine industry began only about forty years ago. Within that short period of time, some 1,700 new wineries have been established in California. Today, there are more than 1,900 wineries in California, most of them making more than one wine, and the price differences are reflected in the styles (you can get a Cabernet Sauvignon wine in any price range from Two Buck Chuck at $1.99 to Harlan Estate at more than $500 a bottle—so how do you choose?). The constant changes in the wine industry through experimentation keep California winemaking in a state of flux.

California Wineries selected by *Wine Spectator* for the 2006 California Wine Experience

Acacia	Crocker & Starr	Hanna
Alcina	Cuvaison Estate	Hanzell
A. P. Vin	Dalla Valle	Harlan
Araujo	Darioush	HdV
Argyle	Delectus	Heitz
David Arthur	Diamond Creek	Hill Climber
Au Bon Climat	Dolce	Paul Hobbs
Aubert	Domaine Alfred	Iron Horse
L'Aventure	Domaine Carneros	J Vineyards & Winery
Bacio Divino	Domaine Chandon	JC Cellars
Barnett	Domain Drouhin Oregon	John Anthony
Beau Vigne	Domaine Serene	Justin
Beaulieu	Dominus	Keller
Bennett Lane	The Donum Estate	Kathryn Kennedy
Bergstrom	Duckhorn	Kenwood
Beringer	DuMOL	Kistler
Betz Family	Dunn	Klinker Brick
BOND	Dutton-Goldfield	Kosta Browne
Brander	Merry Edwards	Kunde
David Bruce	El Molino	Ladera
Buehler	Etude	Lagier Meredith
Buoncristiani	Far Niente	Lail
Burgess	Gary Farrell	Landmark
Byron	Ferrari-Carano	Lang & Reed
Calera	Fisher	Lewis
Carter	Flora Springs	Loring
Caymus	Foley Estate	Luna
Cayuse	Robert Foley	MacRostie
Chalk Hill	Foxen	Margerum
Chalone	Franciscan Oakville	Marimar Estate
Chappellet	Frank Family	Markham
Chateau St. Jean	Gallo Family	Marston Family
Chateau Souverain	Gemstone	Martinelli
Chehalem	Geyser Peak	Mer Soleil
Chimney Rock	Girard	Meridian
Clos du Bois	Gloria Ferrer	Merryvale
Col Solare	Goldeneye	Merus
Cornerstone	Grgich Hills	Peter Michael
Cougar Crest	Groth	Robert Mondavi
Robert Craig	HALL	Morgan

A 40+-YEAR PERSPECTIVE

Number of Bonded Wineries in California

Year	Wineries
1965	232
1970	240
1975	330
1980	508
1985	712
1990	807
1995	944
2000	1,210
2007	1,905

Source: The Wine Institute

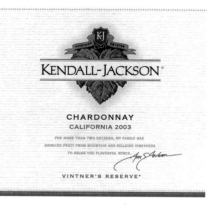

Mueller	Ridge	Steele
Mumm Napa	Roederer Estate	Sterling
Neyers	Rosenblum	Rodney Strong
Nickel & Nickel	Stephen Ross	Tablas Creek
Northstar	Rubicon Estate	Robert Talbott
Novy	Rutherford Hill	Talley
Olabisi	St. Clement	Testarossa
Opus One	St. Francis	TOR
Pahlmeyer	St. Supery	Treana
Paloma	Sanford	Trefethen
Papapietro Perry	Sbragia Family	Trinchero
Patz & Hall	Schrader	Truchard
Penner-Ash	Schramsberg	Turnbull
Peju Province	Sebastiani	Viader
Joseph Phelps	Seghesio	Villa Mt. Eden
Pine Ridge	Selene	Vine Cliff
PlumpJack	Shafer	Vision
Ponzi	Siduri	Whitehall Lane
Pride Mountain	Silver Oak	Williams Selyem
Provenance	W. H. Smith	Robert Young
Quintessa	Snowden	
Quixote	Sonoma-Loeb	
Qupe	Spottswoode	
Ramey	Spring Valley	
Martin Ray	Stag's Leap Wine Cellars	
Raymond	Staglin Family	

What about the prices of California varietal wines?

You can't necessarily equate quality with price. Some excellent varietal wines that are produced in California are well within the budget of the average consumer. On the other hand, some varietals (primarily Chardonnay and Cabernet Sauvignon) may be quite expensive.

As in any market, it is mainly supply and demand that determines price. However, new wineries are affected by start-up costs, which sometimes are reflected in the price of the wine. Older, established wineries, which long ago amortized their investments, are able to keep their prices low when the supply/demand ratio calls for it. Remember, when you're buying California wine, price doesn't always reflect quality.

How do I choose a good California wine?

One of the reasons California produces such a wide variety of wine is that it has so many different climates. Some are as cool as Burgundy, Champagne, and the Rhein, while others are as warm as the Rhône Valley, Portugal, and the southern regions of Italy and Spain. If that's not diverse enough, these wine-growing areas have inner districts with "microclimates," or climates within climates. One of the microclimates (which are among the designated AVAs) in Sonoma County, for example, is the Russian River Valley.

To better understand this concept, let's take a close look at the Rudd label.

State:

California

County:

Sonoma

Viticultural Area (AVA):

Russian River Valley

Vineyard:

Bacigalupi

Winery:

Rudd

California labels tell you everything you need to know about the wine—and more. Here are some quick tips you can use when you scan the shelves at your favorite retailer. The label shown above will serve as an example.

The most important piece of information on the label is the producer's name. In this case, the producer is Rudd.

If the grape variety is on the label, a minimum of 75 percent of the wine must be derived from that grape variety. This label shows that the wine is made from the Chardonnay grape.

If the wine bears a vintage date, 95 percent of the grapes must have been harvested that year.

If the wine is designated "California," then 100 percent of the grapes must have been grown in California.

If the label designates a certain federally recognized viticultural area

THERE ARE 109 AVAs in California. Some of the best known are:

Napa Valley	Livermore Valley
Sonoma Valley	Paso Robles
Russian River Valley	Edna Valley
Alexander Valley	Fiddletown
Dry Creek Valley	Stag's Leap
Los Carneros	Chalk Hill
Anderson Valley	Howell Mountain
Santa Cruz Mountain	

FAMOUS INDIVIDUAL VINEYARDS OF CALIFORNIA

Bacigalupi

Bien Nacido

Dutton Ranch

Durell

Robert Young

Bancroft Ranch

Gravelly Meadow

Martha's Vineyard

McCrea

S.L.V.

To-Kalon

Beckstoffer

Monte Rosso

IF AN INDIVIDUAL vineyard is noted on the label, 95% of the grapes must be from the named vineyard, which must be located within an approved AVA.

THE LEGAL limit for the alcohol content of table wine is 7% to 13.9%, with a 1.5% allowance either way, so long as the allowance doesn't go beyond the legal limit. If the alcohol content of a table wine exceeds 14%, the label must show that. Sparkling wines may be 10% to 13.9%, with the 1.5% allowance.

I'M SURE you'll recognize the names of some of the early European winemakers:

Finland
> Gustave Niebaum (Inglenook)—1879

France
> Paul Masson—1852
> Étienne Thée and Charles LeFranc (Almaden)—1852
> Pierre Mirassou—1854
> Georges de Latour (Beaulieu)—1900

Germany
> Beringer Brothers—1876
> Carl Wente—1883

Ireland
> James Concannon—1883

Italy
> Giuseppe and Pietro Simi—1876
> John Foppiano—1895
> Samuele Sebastiani—1904
> Louis Martini—1922
> Adolph Parducci—1932

"You are never going to stylize the California wines the same way that European wines have been stylized, because we have more freedom to experiment. I value my freedom to make the style of wine I want more than the security of the AOC laws. Laws discourage experimentation."
> —LOUIS MARTINI

(AVA), such as Russian River Valley (as on our sample label), then at least 85 percent of the grapes used to make that wine must have been grown in that location.

The alcohol content is given in percentages. Usually, the higher the percentage of alcohol, the "fuller" the wine will be.

"Produced and bottled by" means that at least 75 percent of the wine was fermented by the winery named on the label.

Some wineries tell you the exact varietal content of the wine, and/or the sugar content of the grapes when they were picked, and/or the amount of residual sugar (to let you know how sweet or dry the wine is).

How is California winemaking different from the European technique?

Many students ask me this, and I can only tell them I'm glad I learned all about the wines of France, Italy, Germany, Spain, and the rest of Europe before I tackled California. European winemaking has established traditions that have remained essentially unchanged for hundreds of years. These practices involve the ways grapes are grown and harvested, and in some cases include winemaking and aging procedures.

In California, there are few traditions, and winemakers are able to take full advantage of modern technology. Furthermore, there is freedom to experiment and create new products. Some of the experimenting the California winemakers do, such as combining different grape varieties to make new styles of wine, is prohibited by some European wine-control laws. Californians thus have greater opportunity to try many new ideas.

Another way in which California winemaking is different from European is that many California wineries carry an entire line of wine. Many of the larger ones produce more than twenty different labels. In Bordeaux, most châteaux produce only one or two wines.

In addition to modern technology and experimentation, you can't ignore the fundamentals of winegrowing: California's rainfall, weather patterns, and soils are very different from those of Europe. The greater abundance of sunshine in California can result in wines with greater alcohol content, ranging on average from 13.5 percent to 14.5 percent, compared to 12 percent to 13 percent in Europe. This higher alcohol content changes the balance and taste of the wines.

EUROWINEMAKING IN CALIFORNIA

Many well-known and highly respected European winemakers have invested in California vineyards to make their own wine. There are more than forty-five California wineries owned by European, Canadian, or Japanese companies. For example:

- *One of the most influential joint ventures matched Baron Philippe de Rothschild, then the owner of Château Mouton-Rothschild in Bordeaux, and Robert Mondavi, of Napa Valley, to produce a wine called Opus One.*
- *The owners of Château Pétrus in Bordeaux, the Moueix family, have vineyards in California. Their wine is a Bordeaux-style blend called Dominus.*
- *Moët & Chandon, which is part of Moët-Hennessy, owns Domaine Chandon in the Napa Valley.*

Other European wineries with operations in California:
- *Roederer has grapes planted in Mendocino County and produces Roederer Estate.*
- *Mumm produces a sparkling wine, called Mumm Cuvée Napa.*
- *Taittinger has its own sparkling wine called Domaine Carneros.*
- *The Spanish sparkling-wine house Codorniu owns a winery called Artesa; and Freixenet owns land in Sonoma County and produces a wine called Gloria Ferrer.*
- *The Torres family of Spain owns a winery called Marimar Torres Estate in Sonoma County.*
- *Frenchman Robert Skalli (Fortant de France) owns more than six thousand acres in Napa Valley and the winery St. Supery.*
- *Tuscan wine producer Piero Antinori owns Atlas Peak winery in Napa.*

What happened when phylloxera returned to the vineyards of California in the 1980s?

In the 1980s the plant louse phylloxera destroyed a good part of the vineyards of California, costing more than a billion dollars in new plantings. Now this may sound strange, but it proved that good can come from bad. So what's the good news?

This time, vineyard owners didn't have to wait to discover a solution; they already knew what they would have to do to replace the dead vines—by replanting with a different rootstock that they knew was resistant to phylloxera. So while the short-term effects were terribly expensive, the long-term effect should be better-quality wine. Why is this?

THE SPORTING LIFE

Athletes who have vineyards in California:
- Tom Seaver (baseball)
- Joe Montana (football)
- John Madden (football)
- Dick Vermeil (football)
- Peggy Fleming (ice skating)
- Arnold Palmer (golf)
- Greg Norman (golf)
- Mario Andretti (auto racing)
- Randy Lewis (auto racing)
- Jeff Gordon (auto racing)

Dominus
Napa Valley
2001
Christian Moueix

ALC.14.1% BY VOL.-750ML
NAPA VALLEY RED WINE
CONTAINS SULFITES - PRODUCED AND BOTTLED BY
DOMINUS ESTATE, YOUNTVILLE, CALIFORNIA, U.S.A.

In the early days of California grape growing, little thought was given to where a specific grape would grow best. Many Chardonnays were planted in climates that were much too warm, and Cabernet Sauvignons were planted in climates that were much too cold.

With the onset of phylloxera, winery owners had a chance to rectify their errors; when replanting, they matched the climate and soil with the best grape variety. Grape growers also had the opportunity to plant different grape clones. But the biggest change was in the planting density of the vines themselves. Traditional spacing used by most wineries was somewhere between four hundred and five hundred vines per acre. Today with the new replanting, it is not uncommon to have more than a thousand vines per acre. Many vineyards have planted more than two thousand per acre.

The bottom line is that if you like California wines now, you'll love them more with time. The quality is already better and the costs are lower—making it a win-win situation for everyone.

THE WHITE WINES OF CALIFORNIA

What is the major white-grape variety grown in California?

The most important white-wine grape grown in California is Chardonnay. This green-skinned (*Vitis vinifera*) grape is considered by many the finest white-grape variety in the world. It is responsible for all the great French white Burgundies, such as Meursault, Chablis, and Puligny-Montrachet. In California, it has been the most successful white grape, yielding a wine of tremendous character and magnificent flavor. The wines are often aged in small oak barrels, increasing their complexity. In the vineyard, yields are fairly low and the grapes command high prices. Chardonnay is always dry, and benefits from aging more than any other American white wine. Superior examples can keep and develop well in the bottle for five years or longer.

Why do some Chardonnays cost more than other varietals?

In addition to everything we've mentioned before, the best wineries age these wines in wood—sometimes for more than a year. Oak barrels have doubled in price over the last five years, averaging eight hundred dollars per barrel. Add to this the cost of the grapes and the length of time before the wine is actually sold, and you can see why the best of the California Chardonnays cost more than twenty-five dollars.

What makes one Chardonnay different from another?

Put it this way: There are many brands of ice cream on the market. They use similar ingredients, but there is only one Ben & Jerry's. The same is true for wine. Among the many things to consider: Is a wine aged in wood or stainless steel? If wood, what type of oak? Was it barrel fermentation? Does the wine undergo a malolactic fermentation? How long does it remain in the barrel (part of the style of the winemaker)? Where do the grapes come from?

The major regions for California Chardonnay are Carneros, Napa, Santa Barbara, and Sonoma.

BEST BETS FOR CALIFORNIA CHARDONNAY

| 1999* | 2001* | 2002* |
| 2003* | 2004* | 2005** |

Note: * signifies exceptional vintage
**signifies extraordinarily good year

What are the other major California white-wine grapes?

Sauvignon Blanc—Sometimes labeled Fumé Blanc. This is one of the grapes used in making the dry white wines of the Graves region of Bordeaux, and the white wines of Sancerre and Pouilly-Fumé in the Loire Valley of France, as well as New Zealand. California Sauvignon Blanc makes one of the best dry white wines in the world. It is sometimes aged in small oak barrels and occasionally blended with the Sémillon grape.

Chenin Blanc—This is one of the most widely planted grapes in the Loire Valley. In California, the grape yields a very attractive, soft, light-bodied wine. It is usually made very dry or semi-sweet; it is a perfect apéritif wine, simple and fruity.

Viognier—One of the major white grapes from the Rhône Valley in France, Viognier thrives in warmer and sunny climates, so it's a perfect grape for the weather conditions in certain areas of California. It has a distinct fragrant bouquet. Not as full-bodied as most Chardonnays, nor as light as most Sauvignon Blancs, it's an excellent food wine.

MALOLACTIC FERMENTATION is a second fermentation that lowers tart malic acids and increases the softer lactic acids, making for a richer style wine. The result is what many wine tasters refer to as a buttery bouquet.

WHY IS SAUVIGNON BLANC often labeled as Fumé Blanc? Robert Mondavi realized that no one was buying Sauvignon Blanc, so he changed its name to Fumé Blanc. Strictly a marketing maneuver—it was still the same wine. Result: Sales took off. Mondavi decided not to trademark the name, allowing anyone to use it (and many producers do).

KEVIN ZRALY'S FAVORITE SAUVIGNON BLANCS

Mantanzas Creek	Caymus
Robert Mondavi	Mason
Simi	Phelps
Silverado	Chalk Hill
Ferrari-Carano	Chateau St. Jean
Dry Creek	Kenwood

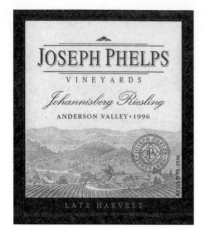

WITH THE success of Italian Pinot Grigio, expect to see many more California wineries producing wines labeled as Pinot Grigio or Pinot Gris.

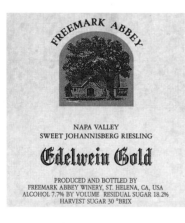

BOTRYTIS CINEREA is a mold that forms on the grapes, known also as "noble rot," which is necessary to make Sauternes and the rich German wines Beerenauslese and Trockenbeerenauslese.

Johannisberg Riesling—The true Riesling responsible for the best German wines of the Rhein and Mosel—and the Alsace wines of France—is also called White Riesling, or just Riesling. This grape produces white wine of distinctive varietal character in every style from bone-dry to very sweet dessert wines, which are often much better by themselves than with dessert. The smell of Riesling at its finest is always lively, fragrant, and both fruity and flowery.

The California Wine Institute has set industry-wide standard definitions for the terms "Late Harvest" and other similar designations that link wine style to picking time. The categories proposed with their associated grape-sugar levels at harvest generally follow the terms established by the German Wine Laws (see pages 87–89).

Early Harvest: Equivalent to a German Kabinett, this term refers to wine made from grapes picked at a maximum of 20° Brix.

Regular or Normal: No specific label designation will be used to connote wines made from fruit of traditional maturity levels, 20° to 24° Brix.

Late Harvest: This term is equivalent to a German Auslese and requires a minimum sugar level of 24° Brix at harvest.

Select Late Harvest: Equivalent to a German Beerenauslese, the sugar-level minimum is 28° Brix.

Special Select Late Harvest: This, the highest maturity-level designation, requires that the grapes be picked at a minimum sugar content of 35° Brix, the same level necessary for a German Trockenbeerenauslese.

CALIFORNIA INGENUITY

Several wineries in California started to market their Late Harvest Riesling with German names such as Trockenbeerenauslese. The German government complained, and this practice was discontinued. One winemaker, though, began marketing his wine as TBA, the abbreviation for Trockenbeerenauslese. Again there was a complaint. This time the winemaker argued his case, saying that TBA was not an abbreviation for Trockenbeerenauslese, but for Totally Botrytis Affected.

Stay tuned for more.

What are the latest trends in the wines of California?

There has been a trend toward wineries specializing in particular grape varieties. Twenty years ago, I would have talked about which wineries in California were the best. Today, I'm more likely to talk about which winery or AVA makes the best Chardonnay; which winery makes the best Sauvignon Blanc; and the same would hold true for the reds, narrowing it down to who makes the best Cabernet Sauvignon, Pinot Noir, Merlot, Zinfandel, or Syrah.

The era of great experimentation with winemaking techniques is slowly coming to an end, and now the winemakers are just making the finest possible wines they can from what they've learned in the 1980s and 1990s. I do expect to see some further experimentation to determine which grape varieties grow best in the various AVAs and microclimates. One of the biggest changes in the last twenty years is that wineries have also become more food conscious in winemaking, adjusting their wine styles to go better with various kinds of food.

Chardonnay and Cabernet Sauvignon remain the two major grape varieties, but much more Syrah is being planted in California. Sauvignon Blanc (also called Fumé Blanc) wines have greatly improved. They're easier to consume young. Although they still don't have the cachet of a Chardonnay, I find them better matched with most foods. However, other white-grape varieties, such as Riesling and Chenin Blanc, aren't meeting with the same success, and they're harder to sell. Still, just to keep it interesting, some winemakers are planting more European varietals, including Italy's Sangiovese, and Grenache and Viognier from France.

Another significant development in California winemaking is the extent to which giant corporations have been buying up small, midsize, and even large wineries. The wine industry, like so many other businesses these days, has been subject to a great deal of consolidation as a result of mergers and acquisitions. See the list on page 66 for some notable examples of this trend.

FOR FURTHER READING

I recommend *The Oxford Companion to the Wines of North America* by Bruce Cass and Jancis Robinson; *The Wine Atlas of California* by James Halliday; *Making Sense of California Wine* and *New California Wine* by Matt Kramer; *California Wine* by James Laube; *American Vintage: The Rise of American Wine* by Paul Lukacs; and *The Wine Atlas of California and the Pacific Northwest* by Bob Thompson. Lovers of gossip will have fun reading *The Far Side of Eden* and *Napa: The Story of an American Eden* by James Conaway. And, of course, *Kevin Zraly's American Wine Guide.*

CALIFORNIA IS BEGINNING to have fun with wine brands: Marilyn Merlot, Lewis Race Car Red, Mia's Playground, and Screw Kappa Napa. Expect more to come.

THE CRYSTAL BALL

Over the next ten years, look for China, India, and Russia to become big importers of American wines.

CALIFORNIA TRENDS THROUGH THE DECADES

'60s and '70s: Jug wines
   (Chablis, Burgundy)
'80s: Varietal wines (Chardonnay, Cabernet
   Sauvignon)
'90s: Varietal location (Cabernet Sauvignon–
   Napa, Pinot Noir–Santa Barbara)
'00s: Specific vineyards for varietals

THE 2004 HARVEST was one of the earliest ever, with some grapes being harvested at the beginning of August.

THE 2005 HARVEST was the largest ever in California.

CONSTELLATION BRANDS, with its many acquisitions—including the Robert Mondavi Winery—edged past E. & J. Gallo Winery to be the biggest wine company in the world.

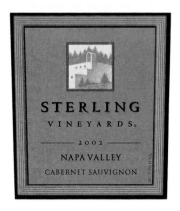

A WINERY BY ANY OTHER NAME

Here, listed by the parent company, is a selection of some well-known wineries and brands.

ALLIED DOMECQ PLC

Atlas Peak • Buena Vista • Callaway Coastal Vineyards • William Hill • Mumm Napa

BERINGER BLASS (FOSTER'S)

Beringer Vineyards • Carmenet • Chateau St. Jean • Chateau Souverian • Etude Wines • Meridian Vineyards • Stags' Leap Winery • St. Clement • Stone Cellars • Windsor

CONSTELLATION BRANDS, INC.

Batavia Wine Cellars • Canandaigua Winery (includes Richards, Arbor Mist, Taylor, J. Roget) • Clos du Bois • Columbia Winery • Covey Run Winery • Dunnewood Vineyards • Estancia Winery • Franciscan Vineyards • Mission Bell Winery (includes Cook's Cribari, Taylor California Cellars, Le Domaine) • Robert Mondavi Winery, Robert Mondavi Napa, Robert Mondavi Private Selection, Opus One, Woodbridge • Mount Veeder Winery • Quintessa • Ravenswood • Ste. Chapelle Winery • Simi Winery • Paul Thomas Winery • Turner Road Vintners (includes Talus, Nathanson Creek, Vendange, Heritage, La Terre)

DIAGEO

Beaulieu Vineyards • BV Coastal • Blossom Hill • The Monterey Vineyard • Painted Hills • Sterling Vineyards

E. & J. GALLO

Anapamu Cellars • Frei Brothers Reserve • Gallo of Sonoma • Indigo Hills • Louis M. Martini • MacMurray Ranch • Marcelina Vineyards • Mirassou Vineyards • Rancho Zabaco • Redwood Creek • Turning Leaf

THE WINE GROUP

Almaden • Colony • Concannon • Corbett Canyon • Franzia • Glen Ellen • Inglenook • Lejon • M.G. Vallejo • Mogen David • Paul Masson • Summit

Three companies—Gallo, Constellation Brands, and the Wine Group make up 60 percent of California wine sales.

Source: Wines & Vines

WINE AND FOOD

MARGRIT BIEVER AND ROBERT MONDAVI: *With Chardonnay: oysters, lobster, a more complex fish with sauce beurre blanc, pheasant salad with truffles. With Sauvignon Blanc: traditional white meat or fish course, sautéed or grilled fish (as long as it isn't an oily fish).*

DAVID STARE *(Dry Creek): With Chardonnay: fresh boiled Dungeness crab cooked in Zatarain's crab boil, a New Orleans–style boil. Serve this with melted butter and a large loaf of sourdough French bread. With Sauvignon Blanc, "I like fresh salmon cooked in almost any manner. Personally, I like to take a whole fresh salmon or salmon steaks and cook them over the barbecue in an aluminum foil pocket. Place the salmon, onion slices, lemon slices, copious quantities of fresh dill, salt, and pepper on aluminum foil and make a pocket. Cook over the barbecue until barely done. Place the salmon in the oven to keep it warm while you take the juices from the aluminum pocket, reduce the juices, strain, and whisk in some plain yogurt. Enjoy!"*

WARREN WINIARSKI *(Stag's Leap Wine Cellars): With Chardonnay: seviche, shellfish, salmon with a light hollandaise sauce.*

JANET TREFETHEN: *With Chardonnay: barbecued whole salmon in a sorrel sauce. With White Riesling: sautéed bay scallops with julienne vegetables.*

RICHARD ARROWOOD: *With Chardonnay: Sonoma Coast Dungeness crab right from the crab pot, with fennel butter as a dipping sauce.*

BO BARRETT *(Chateau Montelena Winery): With Chardonnay: salmon, trout, or abalone, barbecued with olive oil and lemon leaf and slices.*

JACK CAKEBREAD: *"With my 2002 Cakebread Cellars Napa Valley Chardonnay: bruschetta with wild mushrooms, leek and mushroom–stuffed chicken breast, and halibut with caramelized endive and chanterelles."*

ED SBRAGIA *(Beringer Vineyards): With Chardonnay: lobster or salmon with lots of butter.*

U.S. WINE exports have increased from $98 million in 1989 to $951 million in 2007. The top export markets are:

1. United Kingdom
2. Canada
3. Netherlands
4. Japan
5. Germany

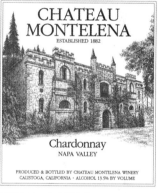

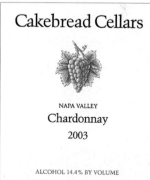

American Wine and Winemaking: A Short History

AMERICANS ARE DRINKING MORE WINE now than ever before. In 2007 Americans consumed more than three gallons of wine per person, the highest ever. The number of American wineries has doubled in the last two decades—to more than four thousand—and, for the first time in American history, all fifty states produce wine.

Still, wine has never been America's favorite beverage. In fact, 40 percent of all Americans don't drink alcohol at all, and another 30 percent don't drink wine. (They prefer beer or distilled spirits.) This leaves only 30 percent of Americans who drink one glass of wine each week, the definition of a "wine drinker" in the United States. In the final analysis, 86 percent of all wine is consumed by 12 percent of the population.

The leading beverage in the United States is soda, with the typical American consuming an average of fifty-four gallons per year. Beer is in second place with thirty-one gallons per person per year. According to Gallup, wine drinking has jumped by more than a third in the United States over the last twenty years, leading an overall increase in alcohol consumption.

Of the 30 percent of U.S. consumers who enjoy wine, the preference is decidedly for American wines: More than 75 percent of all wines consumed by Americans are produced in the United States; the rest are imported. Within the United States, California accounts for 90 percent of wine production. Another 8 percent comes from Washington, New York, and Oregon.

Because American wines are so dominant in the U.S. market, before we move to Class Two it makes sense to pause here and take a detailed look at winemaking in the United States. While we often think of the wine industry as "young" in America, in fact its roots go back some four hundred years.

WE'VE ALL GONE TO LOOK FOR AMERICA:
A PERSONAL WINE JOURNEY

My own love affair with wine began in 1970 when, at the age of nineteen, I visited my first winery, Benmarl, which was—and still is—located in New York's Hudson Valley. That experience struck a nerve. I felt deeply connected to the earth, the grapes, the growers, and, of course, the wine. My soul was stirred; I knew after this first visit that I needed to learn more about wine, winemaking, and wine culture. I continued my early education by seeking out and visiting other wineries in New York State, beginning with wineries in the Hudson Valley and on to those in the Finger Lakes area. I made trip after trip, trekking from vineyard to vineyard.

But that wasn't enough. Each winery I visited provoked an urgent desire to see and learn more about wine. I had taken the first steps of what would become a lifelong journey. I began going to wine tastings as frequently as I could, sampling wines from all over the world. I studied grapes, learning each variety and its characteristics until I was able to identify most of the grapes used in the wines I tasted. And I became obsessive in my study of viticulture and wine tasting. The more I learned, the more I needed to know.

In the early '70s the Finger Lakes district of New York and the North Coast of California were the only two regions in America producing quality wine. I will never forget reading, in 1972, the cover of Time *magazine. The headline read:* AMERICAN WINE: THERE'S GOLD IN THEM THAR GRAPES, *referring to the renaissance of California winemaking. So that summer I took a year off from college and hitchhiked west to California. I stayed for six months, visiting every major winery I'd heard of, tasting the wine and absorbing as much as I could about California wines and wine culture.*

On that trip I discovered that although some California winemakers were committed to producing fine wine, top-quality American wine was still hard to come by. I found only a very few California wines worthy of comparison to the quality European wine I'd tasted. The best California wine was yet to come.

In 1974, after graduating from college, I traveled to Europe to tour the great vineyards of France, Spain, Italy, Germany, and Portugal. I was astounded by the extent to which each country possessed unique wine traditions, producing an impressive variety of superior-quality wines. What struck me most was the difference between the Old World and the New. European wine was like classical music—complex, yet soft and memorable, nurtured

Continued . . .

Time & Life Pictures/Getty Images

IN 2007, wine sales in the United States topped $22 billion.

CALIFORNIA PRODUCES nine-tenths of all wine made in the United States.

THE TOP TEN STATES IN NUMBER OF WINERIES:

1.	California (1905)
2.	Washington (430)
3.	Oregon (290)
4.	New York (218)
5.	Virginia (135)
6.	Texas (124)
7.	Pennsylvania (113)
8.	Ohio (94)
9.	Missouri (92)
10.	Michigan (90)

Source: Wine Business Monthly

CALIFORNIA IS the leading wine consumer in the United States, with more than 42 million cases of wine sold. New York is a distant second, with 19 million cases sold.

THE TOP SEVEN STATES IN WINE CONSUMPTION:

1.	California
2.	New York
3.	Florida
4.	Texas
5.	Illinois
6.	New Jersey
7.	Pennsylvania

and matured over centuries of tradition. American wine, on the other hand, was like rock and roll—young, brash, and new—honoring no rules.

A year after I returned from Europe, fortune smiled on me: I became the first cellar master at Windows on the World—the world-class restaurant atop the newly built World Trade Center in New York City. When we opened, our customers favored European wines, primarily French Bordeaux and Burgundy, which was fine with me. My time abroad had taught me how to taste, what to buy, and how long to age each. However, I was still drawn to American wine, so, in the late '70s, I arranged for a return visit to California. To my delight, this time I found, just as Time had predicted, that California winemakers were beginning to produce more and higher quality wines, some of which were comparable to the finest European wines.

I came back to New York and immediately revised the Windows on the World wine list. My original wine list had been 90 percent French. My new wine list favored American wine by a three-to-one margin. Our customers were cautious, but adventurous enough to taste. Once they sampled and enjoyed the delicious Sonoma Chardonnays and Napa Valley Cabernet Sauvignons, they too became believers. By 1980, American consumers had begun to take California wine seriously.

But California wasn't the only state producing good wine. Other states and regions, such as Oregon, Washington, and Long Island, were developing excellent vineyards and wineries as well. Word on the street quickly spread, and in culinary circles conversation often began with: "Have you tried the Oregon Pinot Noir and Pinot Gris?" followed by, "What about Washington State Cabernet Sauvignon and Long Island Merlot?"

Today, not only do California, New York, Oregon, and Washington State produce great wines, but for the first time in its history, America is becoming well known as a wine-producing nation. Today I can enjoy a meal accompanied by wines from Virginia, Pennsylvania, Texas, or any of the fifty states. This couldn't have happened without Americans rediscovering wine over the last twenty years. In fact, we are now the third-largest wine-consuming nation in the world, with projections indicating that within the next five years the United States will be the top wine consumer worldwide!

Finally, our time has arrived. Thanks to the conviction and determination of American producers, the demands of American consumers, and the savvy of American wine writers, and in spite of the many obstacles that prevented more rapid progress, I can proudly say that many of the best wines in the world are produced in the United States.

So why did it take more than four hundred years for American wines to reach this quality? What about the early days of wine-making in the United States?

The Pilgrims and early pioneers paved the way for American wine. A short time after arriving in America, the early settlers, accustomed to drinking wine with meals, were delighted to find grapevines growing wild. These thrifty, self-reliant colonists thought they had found in this species (*Vitis labrusca*, primarily) a means of producing their own wine, which would end their dependency on costly wine from Europe. The early settlers cultivated the existing local grapevines, harvested the grapes, and made their first American wine. The taste of the new vintage was disappointing, however, for they discovered that wine made from New World grapes possessed an unfamiliar and entirely different flavor than did wine made from European grapes. Undaunted by this first failure and resolute in their determination, cuttings were ordered from Europe of the *Vitis vinifera* vine, which had for centuries produced the finest wines in the world. Soon ships arrived bearing the tender cuttings, and the colonists, having paid scarce, hard-earned money for these new vines, planted and tended them with great care. They were eager to taste their first wine made from European grapes grown in American soil.

What happened?

Nothing but problems! Despite their careful cultivation, few of the European vines thrived. Many died, and those that did survive produced few grapes, whose meager yield resulted in very poor quality wine. Early settlers blamed the cold climate, but today we know that their European vines lacked immunity to the New World's plant diseases and pests. If the colonists had had access to modern methods of pest and disease control, the *Vitis vinifera* grapes would have thrived then, just as they do today. However, for the next

LEIF ERIKSSON, upon discovering North America, named it Vineland. In fact, there are more species of native grapes in North America than on any other continent.

THE FRENCH Huguenots established colonies in Jacksonville, Florida, in 1562 and produced wine using the wild Scuppernong grape. Evidence indicates that there was a flourishing wine industry in 1609 at the site of the early Jamestown settlements. In 2004 an old wine cellar was discovered in Jamestown with an empty bottle dating back to the 17th century.

WILLIAM PENN planted the first vineyard in Pennsylvania in 1683.

EARLY GERMAN immigrants imported Riesling grapes and called their finished wine Hock; the French called their wine Burgundy or Bordeaux; and the Italians borrowed the name "Chianti" for theirs.

The major varieties of wine produced in the U.S. are made from these species:
American: *Vitis labrusca*, such as the Concord, Catawba, and Delaware; and *Vitis rotundifolia*, commonly called Scuppernong.
European: *Vitis vinifera*, such as Riesling, Sauvignon Blanc, Chardonnay, Pinot Noir, Merlot, Cabernet Sauvignon, Zinfandel, and Syrah.
Hybrids: A cross between *vinifera* and a native American species, such as Seyval Blanc, Vidal Blanc, Baco Noir, and Chancellor.

VITIS LABRUSCA, the "slip-skinned" grape, is native to both the Northeast and the Midwest and produces a unique flavor. It is used in making grape juice—the bottled kind you'll find on supermarket shelves. Wine produced from *labrusca* grapes tastes, well, more "grapey" (also described as "foxy") than European wines.

two hundred years every attempt at establishing varieties of *vinifera*—either intact or through crossbreeding with native vines—failed. Left with no choice, growers throughout the Northeast and Midwest returned to planting *Vitis labrusca*, North America's vine, and a small wine industry managed to survive.

Those early Americans never really got used to the taste of this wine, however, and European wine remained the preferred—though high-priced—choice. The failures of these early attempts to establish a wine industry in the United States, along with the high cost of imported wines, resulted in decreasing demand for wine. Gradually, American tastes changed and wine served at mealtime was reserved for special occasions; beer and whiskey had taken over wine's traditional place in American homes.

But when I think of American wines I think of California. How and when did wine arrive in the West?

Wine production in the West began with the Spanish. As the Spanish settlers began pushing northward from Mexico, the Catholic Church followed, and a great era of mission building began. Early missions were more than just churches; they were entire communities conceived as self-sufficient fortifications protecting Spanish colonial interests throughout the Southwest and along the Pacific Coast. Besides growing their own food and making their own clothing, those early settlers also made their own wine, produced primarily for use in the Church. Sacramental wine was especially important in early Church ritual. (Perhaps higher quality wine was an important factor in attracting congregants!) The demand for wine led Padre Junípero Serra to bring *Vitis vinifera* vines—brought to Mexico by the Spaniards—from Mexico to California in 1769. These vines took root, thriving in California due to its moderate climate. The first true California wine industry had been established, albeit on a small scale.

Two events occurred in the mid-1800s that resulted in an explosive growth of quality wine production. The first was the California Gold Rush, in 1849. With a huge rush of immigrants pushing steadily westward, California's population exploded. Along with their hopes of finding treasure, immigrants from Europe and the East Coast brought their winemaking traditions. Unsuccessful in finding gold nuggets, many discovered a different kind of gold: California's grapevines. They cultivated the vines and were soon producing good-quality commercial wine.

The second critical event occurred in 1861, when the governor of California, understanding the importance of viticulture to the state's growing economy,

THE EARLY missionaries established wineries in the southern parts of California. The first commercial winery was established in what we know today as Los Angeles.

THE GRAPE variety the missionaries used to make their sacramental wine was actually called the Mission grape. Unfortunately, it did not have the potential to produce a great wine.

IN 1861, first lady Mary Todd Lincoln served American wines in the White House.

commissioned Count Agoston Haraszthy to select and import classic *Vitis vinifera* cuttings—such as Riesling, Zinfandel, Cabernet Sauvignon, and Chardonnay—from Europe. The count traveled to Europe, returning with more than 100,000 carefully selected vines. Due to the climatic conditions in California, not only did these grape varieties thrive, they also produced good-quality wine! Serious California winemaking began in earnest. It was during that period that the Civil War broke out in the United States, leaving the fledgling wine industry with very little government attention or support. In spite of this, the quality of California wine improved dramatically over the next thirty years.

So, the rest, as they say, is history?

Not at all. In 1863, while California wines were flourishing, European vineyards were in trouble. Phylloxera—an aphid pest native to the East Coast that is very destructive to grape crops—began attacking European vineyards. This infestation, which arrived in Europe on cuttings from native American vines exported for experimental purposes, proved devastating. Over the next two decades, the phylloxera blight destroyed thousands of acres of European vines, severely diminishing European wine production just as demand was rapidly growing.

WHEN ROBERT Louis Stevenson honeymooned in the Napa Valley in 1880, he described the efforts of local vintners to match soil and climate with the best possible varietals. "One corner of land after another . . . this is a failure, that is better, this is best. So bit by bit, they grope about for their Clos de Vougeot and Lafite . . . and the wine is bottled poetry."

Since California was now virtually the only area in the world producing wine made from European grapes, demand for its wines skyrocketed. This helped develop, almost overnight, two huge markets for California wine. The first market clamored for good, inexpensive, yet drinkable wine produced on a mass scale. The second market sought higher-quality wines.

IN 2006, California wine production was 449 million gallons.

California growers responded to both demands, and by 1876 California was producing more than 2.3 million gallons of wine per year, some of remarkable quality. California was, for the moment, the new center of global winemaking.

Unfortunately, in that same year, phylloxera arrived in California and began attacking its vineyards. Once it got there, it spread as rapidly as it had in Europe. With thousands of vines dying, the California wine industry faced financial ruin. To this day, the phylloxera blight remains the most destructive crop epidemic of all time.

What happened next?

THE THREE major wine-producing states in the 19th century were New York, Ohio, and Missouri.

Luckily, other states had continued producing wine made from *labrusca* vines, and American wine production didn't grind to a complete halt. Meanwhile, after years of research, European winemakers finally found a defense against

FORTY DIFFERENT American wineries won medals at the 1900 Paris Exposition, including those from California, New Jersey, New York, Ohio, and Virginia.

IN 1920 there were more than 700 wineries in California. By the end of Prohibition there were 160.

"Once, during Prohibition, I was forced to live for days on nothing but food and water."
—W. C. FIELDS

WELCH'S GRAPE JUICE

Welch's grape juice has been around since the late 1800s. It was originally produced by staunch Prohibitionists and labeled "Dr. Welch's Unfermented Wine." In 1892 it was renamed "Welch's Grape Juice" and was successfully launched at the Columbian Exposition in Chicago in 1893.

the pernicious phylloxera aphid. They were the first to successfully graft *Vitis vinifera* vines onto the rootstock of *labrusca* vines (which were immune to the phylloxera), rescuing their wine industry.

Americans followed, and the California wine industry not only recovered but began producing better-quality wines than ever before. By the late 1800s, California wines were winning medals in international competition, gaining the respect and admiration of the world. And it took only three hundred years!

PROHIBITION: YET ANOTHER SETBACK

In 1920, the Eighteenth Amendment to the United States Constitution was enacted, creating yet another setback for the American wine industry. The National Prohibition Act, also known as the Volstead Act, prohibited the manufacture, sale, transportation, importation, exportation, delivery, or possession of intoxicating liquors for beverage purposes. Prohibition nearly destroyed what had become a thriving and national industry.

If Prohibition had lasted only four or five years, its impact on the wine industry might have been negligible. Unfortunately, it continued for thirteen years, during which grapes became important to the economy of the criminal world. One of the loopholes in the Volstead Act allowed for the manufacture and sale of sacramental wine, medicinal wines for sale by pharmacists with a doctor's prescription, and medicinal wine tonics (fortified wines) sold without prescription. Perhaps more important, Prohibition allowed anyone to produce up to two hundred gallons yearly of fruit juice or cider. The fruit juice, which was sometimes made into concentrate, was ideal for making wine. Some of this yield found its way to bootleggers throughout America who did just that. But not for long, because the government stepped in and banned the sale of grape juice, preventing illegal wine production. Vineyards stopped being planted, and the American wine industry came to a halt.

The Roaring Twenties fostered more beer and distilled-spirit drinkers than wine drinkers, because the raw materials used in production were easier to come by. But fortified wine, or medicinal wine tonic—containing about 20 percent alcohol, which makes it more like a distilled spirit than regular wine—was still available and became America's number one wine. American wine was soon popular more for its effect than its taste; in fact, the word *wino* came into use during the Depression from the name given to those unfortunate souls who turned to fortified wine to forget their troubles.

Prohibition came to a close in 1933, but its impact would be felt for decades. By its end, Americans had lost interest in quality wine. During Prohibition, thousands of acres of valuable grapes around the country had been plowed under. Wineries nationwide shut down and the winemaking industry dwindled to a handful of survivors, mostly in California and New York. Many growers on the East Coast returned to producing grape juice—the ideal use for the American *labrusca* grape.

The federal government, in repealing Prohibition, empowered states to legislate the sale and transportation of alcohol. Some states handed control to counties and, occasionally, even municipalities—a tradition that continues today, varying from state to state and often from county to county.

HARD TIMES FOR WINE: 1933–68

Although Prohibition was devastating to the majority of American wine producers, some endured by making sacramental wines. Beringer, Beaulieu, and the Christian Brothers are a few of the wineries that managed to survive this dry time. Since these wineries didn't have to interrupt production during Prohibition, they had a jump on those that had to start all over again in 1933.

From 1933 to 1968, grape growers and winemakers had little more than personal incentive to produce any wine of quality. Jug wines, which got their name from the containers in which they were bottled, were inexpensive, nondescript, and mass produced. A few wineries, notably in California, were producing some good wines, but the majority of American wines produced during this period were ordinary.

SOME OF THE dilemmas facing winemakers after Prohibition:
  Locate on the East Coast or the West Coast?
  Make sweet wine or dry wine?
  Make high alcohol wine or low alcohol wine?
  Make inexpensive bulk wine or premium wine?

BEST-SELLING wineries, from 1933 to 1968, were Almaden, Gallo, and Paul Masson.

IN 1933, more than 60% of wine sold in the United States contained more than 20% alcohol.

IN 1930, there were 188,000 acres of vineyards in California. Today, there are over half a million acres.

IN 1960, THERE were only 10 wineries in Napa Valley.

THE BEST-KNOWN wineries of California in the 1960s and 1970s:

Almaden	Korbel
Beaulieu	Krug
Beringer	Martini
Concannon	Paul Masson
Inglenook	Wente

IN THE early 1970s, Chenin Blanc was the best-selling white wine and Zinfandel the best-selling red.

IN 2007, there were more than 4,000 wineries in the United States, up from 580 in 1975. Ninety-nine percent are small and family owned.

THE MAJORITY (57%) of American wineries are located in states other than California.

THE RENAISSANCE OF AMERICAN WINE

I can't say when, exactly, the American wine renaissance began, but let's start in 1968, when, for the first time since Prohibition, table wines—wines with alcohol content between 7 and 14 percent—outsold fortified wines. Although American wines were improving, consumers still believed the best wines were made in Europe, especially France.

In the mid-sixties and early seventies, a small group of dedicated winemakers, determined that California produce wines equal to the finest of France, began concentrating on making high-quality wine. Their early wines, though not world class, demonstrated potential and began attracting the attention of astute wine writers and wine enthusiasts around the country.

As they continued to improve their product, these same winemakers began to realize that in order to market their wine successfully they had to find a way to differentiate their quality wines from California's mass-produced wines—which had such generic names as Burgundy, Chablis, or Chianti—and to ally their wines, at least in the minds of wine buyers and consumers, with European wines. Their solution was brilliant: They chose to label their best wines by varietal.

Varietal designation calls the wine by the name of the predominant grape used to produce it: Chardonnay, Cabernet Sauvignon, Pinot Noir, etc. The savvy consumer learned that a wine labeled Chardonnay would have the general characteristics of any wine made from that grape. This made wine buying easier for both wine buyers and sellers.

Varietal labeling quickly spread throughout the industry and became so successful that, in the eighties, varietal designation became an American industry standard, forcing the federal government to revise its labeling regulations.

Today, varietal labeling is the norm for the highest-quality American wines, has been adopted by many other countries, and has helped bring worldwide attention to California wine. While California still produces 90 percent of American wine, its success has inspired winemakers in other areas of the United States to refocus on producing high-quality wines.

How does all this help me select and enjoy American wines?

This short history should help you understand and appreciate the trials and tribulations American grape growers and winemakers have endured for centuries. To buy American wines intelligently means having knowledge about and familiarity with each state whose wine you're interested in buying, as well as the regions within the state. Some states—or even regions within a state—may specialize in white wine, others in red, and, going further, there are even regions that specialize in wine made from a specific grape variety. Therefore, it is helpful to know the defined grape-growing areas within each state or region, the American Viticultural Areas (AVAs).

WINERIES IN the United States are opening at the rate of 300 per year.

What is an AVA?

An AVA, or American Viticultural Area, is a specific grape-growing area within a state or a region recognized by and registered with the federal government. AVA designation began in the 1980s and is a system styled after the European regional system. In France, Bordeaux and Burgundy are strictly enforced regional appellations (marked *Appellation d'Origine Contrôlée,* or AOC); in Italy, Tuscany and Piedmont are recognized as zones (marked *Denominazione di Origine Controllata,* or DOC). The Napa Valley, for example, is a defined viticultural area in the state of California. Columbia Valley is an AVA located in

2001
COLUMBIA WINERY
COLUMBIA VALLEY
SYRAH

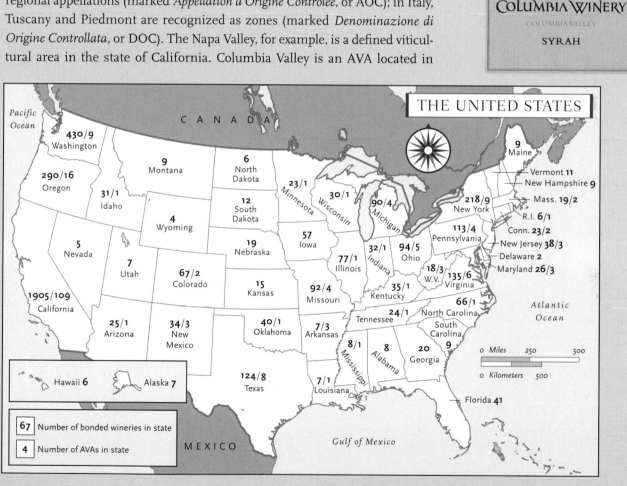

THE UNITED STATES

Pacific Ocean

CANADA

430/9 Washington

9 Maine

290/16 Oregon

9 Montana

6 North Dakota

Minnesota

Wisconsin

23/1

30/1

90/4 Michigan

Vermont **11**
New Hampshire **9**

218/9 New York

Mass. **19/2**

31/1 Idaho

12 South Dakota

57 Iowa

32/1 Indiana

94/5 Ohio

113/4 Pennsylvania

R.I. **6/1**
Conn. **23/2**
New Jersey **38/3**
Delaware **2**
Maryland **26/3**

4 Wyoming

19 Nebraska

77/1 Illinois

5 Nevada

7 Utah

67/2 Colorado

15 Kansas

92/4 Missouri

35/1 Kentucky

18/3 W.V.

135/6 Virginia

66/1 North Carolina

Atlantic Ocean

1905/109 California

25/1 Arizona

34/3 New Mexico

40/1 Oklahoma

7/3 Arkansas

Tennessee **24/1**

South Carolina

9

o Miles   250   500

o Kilometers   500

Hawaii **6**   Alaska **7**

124/8 Texas

8/1 Mississippi

8 Alabama

20 Georgia

7/1 Louisiana

Florida **41**

Gulf of Mexico

MEXICO

67 Number of bonded wineries in state

4 Number of AVAs in state

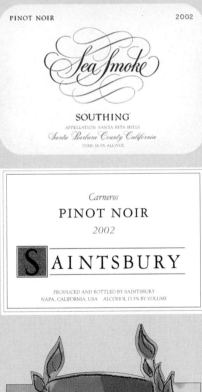

Washington State; both Oregon's Willamette Valley and New York's Finger Lakes district are similarly identified.

Vintners are discovering, as their European counterparts did years ago, which grapes grow best in which particular soils and climatic conditions. There are nearly two hundred viticultural areas in the United States, 109 of which are located in California. I believe the AVA concept is important to wine buying and will continue to be so as individual AVAs become known for certain grape varieties or wine styles. If an AVA is listed on the label, at least 85 percent of the grapes must come from that region.

For example, let's look at Napa Valley, which is probably the best-known AVA in the United States, renowned for its Cabernet Sauvignon. Within Napa, there is a smaller inner district called Carneros, which has a cooler climate. Since Chardonnay and Pinot Noir need a cooler growing season to mature properly, these grape varieties are especially suited to that AVA. In New York, the Finger Lakes region is noted for Riesling. And those of you who have seen the movie Sideways know that Santa Barbara is a great place for Pinot Noir.

Although not necessarily a guarantee of quality, an AVA designation identifies a specific area well known and established for its wine. It is a point of reference for winemakers and consumers. A wine can be better understood by its provenance, or where it came from. The more knowledge you have about a wine's origin, by region and grape, the easier it is to buy even unknown brands with confidence.

How do I choose an American wine?

The essence of this book is to simplify wine for those who are buying or selling it. Do you prefer lighter or heavier wines; red wine or white; sweet wine or dry? You must learn the general characteristics of the major white and red grapes (see pages 15 and 93). Understanding these fundamental differences makes selecting an appropriate wine less difficult, as they help define the wine's style—and selecting the style of wine you want is the first decision you'll need to make.

Next, determine your price range. Are you looking for a nice, everyday wine for under ten dollars? Or are you in the market for a twenty-five- or one-hundred-dollar wine? Set your limit and stick to it. You'll find the style of wine you're looking for at almost any price.

Finally, learn how to read the label (see page 59). Some of the highest-quality wines in the United States come from individual vineyards. The general rule is: The more specific the label, the better the quality of wine. All the important information about any American wine appears on the label. Since the federal government controls wine labeling and has established standards, all American wine labels, regardless of where in the United States the wine was produced, contain essential information that conforms to national standards. This standardization can assist you in making informed decisions about the wine you're about to purchase or pour.

Is that all I need to know?

There is more: An even higher quality of wine is now given a "proprietary" name.

The most recent worldwide trend is to ignore all existing standards by giving the highest-quality wines a proprietary name. A proprietary name helps high-end wineries differentiate their best wines from other wines from the same AVA, from similar varietals, and even from their own other offerings. In the United States, many of these proprietary wines fall under the category called Meritage (see page 142). Some examples of American proprietary wines are Dominus, Opus One, and Rubicon.

Federal laws governing standards and labels are another reason select wineries are increasingly using proprietary names. Federal law mandates, for example, that if a label lists a varietal, at least 75 percent of the grapes used to make the wine must be of that varietal.

Imagine a talented, innovative winemaker in the Columbia Valley region of Washington State. This winemaker is determined to produce an outstanding, full-bodied Bordeaux-style wine consisting of 60 percent Cabernet Sauvignon blended with several other grapes. Our ambitious winemaker has used his best soil for the vines, nurturing them with care and love. He has invested considerable time and labor to produce a really great wine: a wine suitable for aging that will be ready to drink in five years—but will be even better in ten.

After five years, our winemaker tastes the fruits of his labor and voilà! It is delicious, with all the promise of a truly outstanding wine. But how does he distinguish this wine; how can he attract buyers willing to pay a premium price for an unknown wine? He can't label it Cabernet Sauvignon, because less than 75 percent of the grapes used are of that type. For this reason, many producers of fine wine are beginning to use proprietary names. It's indicative of the healthy state of the American wine industry as well. More and more winemakers are turning out better and better wines, and the very best is yet to come!

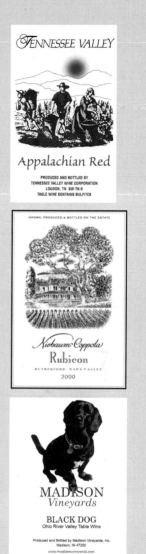

Questions for Class Two: The Wines of Washington, Oregon, and New York; The White Wines of California

The White Wines of Germany

GRAPE VARIETIES • THE STYLE OF GERMAN WINES •

QUALITÄTSWEIN MIT PRÄDIKAT •

UNDERSTANDING GERMAN WINE LABELS • TRENDS

About German Wines

THERE HAS been a 10% increase in the vineyards planted over the last 10 years.

FRANCE PRODUCES 10 times as much wine as Germany.

IN GERMANY, 100,000 grape growers cultivate nearly 270,000 acres of vines, meaning the average holding per grower is 2.7 acres.

Bᴇғᴏʀᴇ ᴡᴇ ʙᴇɢɪɴ ᴏᴜʀ sᴛᴜᴅʏ of the white wines of Germany, tell me this: Have you memorized the 7 Grands Crus of Chablis, the 32 Grands Crus of the Côte d'Or, and the 391 different wineries of the Napa Valley? I hope you have, so you can begin to memorize the more than 1,400 wine villages and 2,600-plus vineyards of Germany. No problem, right? What's 4,000 simple little names?

Actually, if you were to have studied German wines before 1971, you would have had thirty thousand different names to remember. There used to be very small parcels of land owned by an assortment of different people; that's why so many names were involved.

In an effort to make German wines less confusing, the government stepped in and passed a law in 1971. The new ruling stated that a vineyard must encompass at least twelve and a half acres of land. This law cut the list of vineyard names considerably, but it increased the number of owners.

Germany produces only 2 or 3 percent of the world's wines. (Beer, remember, is the national beverage.) And what wines it does produce depends largely on the weather. Why is this? Well, look at where the wines are geographically. Germany is the northernmost country in which vines can grow. And 80 percent of the

GERMANY

North Sea

Hamburg

Berlin ★

Rhein River

Bonn

MOSEL

RHEINGAU

Frankfurt

RHEINHESSEN

Mosel River

PFALZ

Rhein River

Munich

0 Miles   100   200

0 Kilometers   200

vineyards are located on hilly slopes. Germans can forget about mechanical harvesting.

The following chart should help give you a better idea of the hilly conditions vintners must contend with in order to grow grapes and produce German wines.

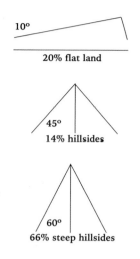

10°
20% flat land

45°
14% hillsides

60°
66% steep hillsides

What are the most important grape varieties?

Riesling—This is the most widely planted and the best grape variety produced in Germany. If you don't see the name "Riesling" on the label, then there's probably very little, if any, Riesling grape in the wine. And remember, if the label gives the grape variety, then there must be at least 85 percent of that grape in the wine, according to German law. Of the grapes planted in Germany, 23 percent are Riesling.

Silvaner—This is another grape variety, and it accounts for 7 percent of Germany's wines.

Müller-Thurgau—A cross between two grapes (Riesling and Chasselas).

What are the main winemaking regions of Germany?

There are thirteen winemaking regions. Do you have to commit them all to memory like the hundreds of other names I've mentioned in the book so far?

Absolutely not. Why should you worry about all thirteen when you only need to be familiar with four?

One of the reasons I emphasize these regions above the others is that in the United States you rarely see wine from the other German wine-growing regions. The other reason to look closely at these regions is that they produce the best German wines. They are:

RHEINHESSEN

RHEINGAU

MOSEL

PFALZ
(UNTIL 1992 KNOWN
AS RHEINPFALZ)

GERMANY IS RIESLING COUNTRY

Germany has more than 50,000 acres of Riesling and no other country comes close. Australia is next, with 10,000 acres, and France has 8,000.

Germany has been growing the Riesling grape since 1435.

THE "OTHER" NINE WINE REGIONS OF GERMANY ARE:

Ahr
Baden
Franken
Hessische Bergstrasse
Mittelrhein
Nahe
Saale-Unstrut
Sachsen
Württemberg

What's the style of German wines?

A balance of sweetness with acidity and low alcohol. Remember the equation:

$$\text{Sugar} + \text{Yeast} = \text{Alcohol} + \text{Carbon Dioxide (CO}_2\text{)}$$

Where does the sugar come from? The sun! If you have a good year, and your vines are on a southerly slope, you'll get a lot of sun, and therefore the right sugar content to produce a good wine. Many times, however, the winemakers aren't so fortunate and they don't have enough sun to ripen the grapes. The result: higher acidity and lower alcohol. To compensate for this, some winemakers may add sugar to the must before fermentation to increase the amount of alcohol. As mentioned before, this process is called chaptalization. (Note: Chaptalization is not permitted for higher-quality German wines.)

The three basic styles of German wine are:

Trocken—dry
Halbtrocken—medium-dry
Fruity—semidry to very sweet

A NOTE ON SÜSSRESERVE

A common misconception about German wine is that fermentation stops and the remaining residual sugar gives the wine its sweetness naturally. On the contrary, some wines are fermented dry. Many German winemakers hold back a certain percentage of unfermented grape juice from the same vineyards, the same varietal, and the same sweetness level. This Süssreserve contains all the natural sugar and it's added back to the wine after fermentation. The finest estates do not use the Süssreserve method, but rely on stopping the fermentation to achieve their style.

What are the ripeness levels of German wine?

As a result of the German law of 1971, there are two main categories, Tafelwein and Qualitätswein.

Tafelwein—Literally, "table wine." The lowest designation given to a wine grown in Germany, it never carries the vineyard name. It is rarely seen in the United States.

Qualitätswein—Literally, "quality wine," of which there are two types.

1. *Qualitätswein bestimmter Anbaugebiete:* QbA indicates a quality wine that comes from one of the thirteen specified regions.

2. *Präedikatswein:* This is quality wine with distinction—the good stuff. These wines may not be chaptalized: The winemaker is not permitted to add sugar. In ascending order of quality, price, and ripeness at harvest, here are the Präedikatswein levels:

Kabinett—Light, semidry wines made from normally ripened grapes. Cost: $15–$25.

Spätlese—Breaking up the word, *spät* means "late" and *lese* means "picking." Put them together and you have "late picking." That's exactly what this medium-style wine is made of—grapes that were picked after the normal harvest. The extra days of sun give the wine more body and a more intense flavor. Cost: $20–$35.

Auslese—Translated as "out picked," this means that the grapes are selectively picked out from particularly ripe bunches, which yields a medium to fuller style wine. You probably do the same thing in your own garden if you grow tomatoes: You pick out the especially ripe ones, leaving the others on the vine. Cost: $25–$50.

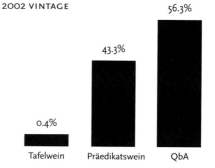

2002 VINTAGE

Tafelwein 0.4% — Präedikatswein 43.3% — QbA 56.3%

FIFTY YEARS AGO, most German wines were dry and very acidic. Even in the finer restaurants, you'd be offered a spoonful of sugar with a German wine to balance the acidity.

GERMAN WINES tend to be 8% to 10% alcohol, compared to an average 11% to 13% for French wines.

OF MOSEL wines, 80% are made from the Riesling grape, while 82% of Rheingau wines are made from Riesling.

JOSEF LEITZ

WEINGUT

Rüdesheimer Berg Roseneck
Riesling Spätlese

Qualitätswein mit Prädikat
Erzeugerabfüllung
alc. 8,5% vol.   A. P. Nr. 24079 018 98   ℮ 750 ml
Product of Germany

D-65385 RÜDESHEIM AM RHEIN
RHEINGAU

GIVEN GOOD WEATHER, the longer the grapes remain on the vine, the sweeter they become—but the winemaker takes a risk when he does this because all could be lost in the event of bad weather.

TODAY, MOST German wines, including Beerenauslese and Trockenbeerenauslese, are bottled in spring and early summer. Many no longer receive additional cask or tank maturation, because it has been discovered that this extra barrel aging destroys the fruit.

IN 1921, the first Trockenbeerenauslese was made in the Mosel region.

THE SPÄTLESE RIDER: THE FIRST LATE-HARVEST WINE

The story goes that at the vineyards of Schloss Johannisberg, the monks were not allowed to pick the grapes until the Abbot of Fulda gave his permission. During the harvest of 1775, the abbot was away attending a synod. That year the grapes were ripening early and some of them had started to rot on the vine. The monks, becoming concerned, dispatched a rider to ask the abbot's permission to pick the grapes. By the time the rider returned, the monks believed all was lost, but they went ahead with the harvest anyway. To their amazement, the wine was one of the best they had ever tasted. That was the beginning of Spätlese-style wines.

Beerenauslese—Breaking the word down, you get *beeren*, or "berries," *aus*, or "out," and *lese*, or "picking." Quite simply (and don't let the bigger names fool you), these are berries (grapes) that are picked out individually. These luscious grapes are used to create the rich dessert wines for which Germany is known. Beerenauslese is usually made only two or three times every ten years. It's not unheard of for a good Beerenauslese to cost up to $250.

Trockenbeerenauslese—A step above the Beerenauslese, but these grapes are dried (*trocken*), so they're more like raisins. These "raisinated" grapes produce the richest, sweetest, honeylike wine—and the most expensive.

WHAT IS *BOTRYTIS CINEREA*?

Botrytis cinerea, known as Edelfäule in German, is a mold that (under special conditions) attacks grapes, as was described in the section on Sauternes. I say "special" because this "noble rot" is instrumental in the production of Beerenauslese and Trockenbeerenauslese.

Noble rot occurs late in the growing season when the nights are cool and heavy with dew, the mornings have fog, and the days are warm. When noble rot attacks the grapes, they begin to shrivel and the water evaporates, leaving concentrated sugar. (Remember, 86 percent of wine is water.) Grapes affected by this mold may not look very appealing, but don't let looks deceive you: The proof is in the wine.

Eiswein—A very rare, sweet, concentrated wine made from frozen grapes left on the vine. They're pressed while still frozen. According to Germany's 1971 rules for winemaking, this wine must now be made from grapes that are at least ripe enough to make a Beerenauslese.

What's the difference between a $100 Beerenauslese and a $200 Beerenauslese (besides a hundred bucks)?

The major difference is the grapes. The $100 bottle is probably made from Müller-Thurgau grapes or Silvaner, while the $200 bottle is from Riesling. In addition, the region the wine comes from will, in part, determine its quality. Traditionally, the best Beerenauslese and Trockenbeerenauslese come from the Rhein or the Mosel.

Quality is higher in wine when:

1. The wine is produced from low yields.

2. The grapes come from great vineyards.

3. The wine is produced by great winemakers.

4. The grapes were grown in a great climate or are from a great vintage.

When I'm ordering a German wine in a restaurant or shopping at my local retailer, what should I look for?

The first thing I would make sure of is that it comes from one of the four major regions. These regions are the Mosel-Saar-Ruwer, Rheinhessen, Rheingau, and Pfalz, which, in my opinion, are the most important quality wine–producing regions in all of Germany.

Next, look to see if the wine is made from the Riesling grape. Anyone who studies and enjoys German wines finds that Riesling shows the best-tasting characteristics. Riesling on the label is a mark of quality.

Also, be aware of the vintage. It's important, especially with German wines, to know if the wine was made in a good year.

Finally, the most important consideration is to buy from a reputable grower or producer.

ONE QUICK way to tell the difference between a Rhein and a Mosel wine on sight is to look at the bottle. Rhein wine comes in a brown bottle, Mosel in a green bottle.

What's the difference between Rhein and Mosel wines?

Rhein wines generally have more body than do Mosels. Mosels are usually higher in acidity and lower in alcohol than are Rhein wines. Mosels show more autumn fruits like apples, pears, and quince, while Rhein wines show more summer fruits like apricots, peaches, and nectarines.

Some important villages to look for:

Rheingau: Eltville, Erbach, Rüdesheim, Rauenthal, Hochheim, Johannisberg

Mosel (formerly Mosel-Saar-Ruwer): Erden, Piesport, Bernkastel, Graach, Ürzig, Brauneberg, Wehlen

Rheinhessen: Oppenheim, Nackenheim, Nierstein

Pfalz: Deidesheim, Forst, Wachenheim, Ruppertsberg, Dürkheimer

WHOSE VINEYARD IS IT, ANYWAY?

Piesporter Goldtröpfchen—350 owners
Wehlener Sonnenuhr—250 owners
Brauneberger Juffer—180 owners

ALL QUALITÄTSWEIN and Präedikatswein must pass a test by an official laboratory and tasting panel to be given an official number, prior to the wine's release to the trade.

IMPRESS YOUR friends with this trivia:
A.P. Nr. 2 602 041 008 02
- **2** = the government referral office or testing station
- **602** = location code of bottler
- **041** = bottler ID number
- **008** = bottle lot
- **02** = the year the wine was tasted by the board

Can you take the mystery out of reading German wine labels?

German wine labels give you plenty of information. For example, see the label at the bottom of page 90.

Joh. Jos. Christoffel Erben is the producer.

Mosel is the region of the wine's origin. Note that the region is one of the big four we discussed earlier in this chapter.

2001 is the year the grapes were harvested.

Ürzig is the town and **Würzgarten** is the vineyard from which the grapes originate. The Germans add the suffix "er" to make *Ürziger*, just as a person from New York is called a New Yorker.

Riesling is the grape variety. Therefore, this wine is at least 85 percent Riesling.

Auslese is the ripeness level, in this case from bunches of overripe grapes.

Qualitätswein mit Prädikat is the quality level of the wine.

A.P. Nr. 2 602 041 008 02 is the official testing number—proof that the wine was tasted by a panel of tasters and passed the strict quality standards required by the government.

Gutsabfüllung means "estate-bottled."

AS OF THE 2007 VINTAGE, Qualitätswein mit Prädikat has been replaced by Präedikatswein.

What are the recent trends in the white wines of Germany?

Germany is producing higher quality wines than ever before, and American interest in these wines has increased at the same time.

I feel the lighter-style Trocken (dry), Halbtrocken (medium-dry), Kabinetts, and even Spätleses are wines that can be easily served as an apéritif, or with very light food and also grilled food, and in particular with spicy or Pacific Rim cuisines. If you haven't had a German wine in a long time, the 2001 through 2005 are spectacular vintages that show the greatness of what German white wines are all about.

And the great news is that those hard-to-read gothic script German wine labels have become more user-friendly—easier to read, with more modern designs.

THE 2003 HARVEST was one of the earliest in decades.

AS GERMAN winemakers say, "A hundred days of sun will make a good wine, but 120 days of sun will make a great wine."

PAST GREAT vintages of Beerenauslese and Trockenbeerenauslese: 1985, 1988, 1989, 1990, 1996.

FOR FURTHER READING, the *Gault-Millan Guide to German Wine* by Armin Diel and Joel Payne.

BEST BETS FOR RECENT VINTAGES IN GERMANY

2001* 2002* 2003* 2004* 2005** 2006 2007

Note: * signifies exceptional vintage   ** signifies extraordinary vintage

WINE AND FOOD

RAINER LINGENFELDER *(Weingut Lingenfelder Estate, Pfalz): With Riesling Spätlese Halbtrocken, "We have a tradition of cooking freshwater fish that come from a number of small creeks in the Palatinate forest, so my personal choice would be trout, either herbed with thyme, basil, parsley, and onion and cooked in wine, or smoked with a bit of horseradish. We find it to be a very versatile wine, a very good match with a whole range of white meat. Pork is traditional in the Palatinate region, as are chicken and goose dishes."*

JOHANNES SELBACH *(Selbach-Oster, Mosel): "What kinds of food do we have with Riesling Spätlese? Anything we like! This may sound funny, but there's a wide variety of food that goes very well with—and this is the key—a fruity, only moderately sweet, well-balanced Riesling Spätlese. Start with mild curries and sesame- or ginger-flavored not-too-spicy dishes. Or try either*

gravlax or smoked salmon. You can even have Riesling Spätlese with a green salad in a balsamic vinaigrette, preferably with a touch of raspberry, as long as the dressing is not too vinegary. Many people avoid pairing wine with a salad, but it works beautifully.

"For haute cuisine, fresh duck or goose liver lightly sautéed in its own juice, or veal sweetbreads in a rich sauce. Also salads with fresh greens, fresh fruit, and fresh seafood marinated in lime or lemon juice or balsamic vinegar.

"With an old ripe Spätlese: roast venison, dishes with cream sauces, and any white-meat dish stuffed with or accompanied by fruit. It is also delicious with fresh fruit itself or as an apéritif."

With Riesling Spätlese Halbtrocken: "This is a food-friendly wine, but the first thing that comes to mind is fresh seafood and fresh fish. Also wonderful with salads with a mild vinaigrette, and with a course that's often difficult to match: cream soups. If we don't know exactly what to drink with a particular food, Spätlese Halbtrocken is usually the safe bet.

"It may be too obvious to say foie gras with Eiswein, but it is a classic."

Red Grapes of the World

CLASSES FOUR THROUGH SEVEN will delve into the great red wines of the world. Just as you studied the major white grape varieties before the classes on white wines, as you begin this journey you should understand the major red grape varieties and where in the world they produce the best wines.

In Class Four, I start with a list of what I consider to be the major red-wine grapes, ranked from lightest to fullest-bodied style, along with the region or country in which the grape grows best. By looking at this chart, not only will you get an idea of the style of the wine, but also a feeling for gradations of weight, color, tannin, and ageability.

THERE ARE hundreds of different red-wine grapes planted throughout the world. California alone grows 31 different red-wine grape varieties.

IN GENERAL, the lighter the color, the more perceived acidity.

ONCE YOU have become acquainted with these major red-wine grapes, you may wish to explore the following:

TEXTURE	GRAPES	TANNIN LEVEL	WHERE THEY GROW BEST	COLOR LEVEL	AGEABILITY
Light		Low		Lighter	Drink young
	Gamay		Beaujolais, France		
	Pinot Noir		Burgundy, France; Champagne, France; California; Oregon		
	Tempranillo		Rioja, Spain		
	Sangiovese		Tuscany, Italy		
	Merlot		Bordeaux, France; California; Washington State		
	Zinfandel		California		
	Cabernet Sauvignon		Bordeaux, France; California; Chile		
	Nebbiolo		Piedmont, Italy		
	Syrah/Shiraz		Rhône Valley, France; Australia; California		
Full-bodied		High		Deeper	Wine to age

GRAPES	WHERE THEY GROW BEST
Barbera	Italy
Dolcetto	Italy
Cabernet Franc	Loire Valley and Bordeaux, France
Grenache/ Garnacha	Rhône Valley, France Spain
Malbec	Bordeaux and Cahors, France; Argentina

To put this chart together is extremely challenging, given all the variables that go into making wine and the many different styles that can be produced. Remember, there are always exceptions to the rule, just as there are other countries and wine regions not listed here that produce world-class wine from some of the red grapes shown. You'll begin to see this for yourself if you do your homework and taste a lot of different wines. Good luck!

Questions for Class Three:
The White Wines of Germany

12. What are the two main categories of Qualitätswein?

13. Name the levels of ripeness of Praedikatswein.

14. What does the word *Spätlese* mean in English?

15. What is the average range of alcohol levels of German wines?

16. What is Eiswein?

17. What is *Botrytis Cinerea* (Edelfäule)?

18. Match the region to the village.

    a. Rheingau   ___Oppenheim

    b. Mosel   ___Rüdesheim

    c. Rheinhessen   ___Bernkastel

    d. Pfalz   ___Piesport

                          ___Johannisberg

                          ___Deidesheim

                          ___Nierstein

19. What does the German term *Gutsabfüllung* indicate?

20. What does the German term *Trocken* say about the style of the wine?

The Red Wines of Burgundy and the Rhône Valley

THE RED WINES OF BURGUNDY · WINE-PRODUCING AREAS · BEAUJOLAIS ·

CÔTE CHÂLONNAISE · CÔTE D'ÔR · CÔTE DE BEAUNE ·

CÔTE DE NUITS · RHÔNE VALLEY

The Red Wines of Burgundy

Now we're getting into a whole new experience in wines—the reds. The complexity, nuances, flavors, and length of taste in red wines evoke a sense of excitement for me and my students, and we tend to concentrate more when we taste red wines.

2,000 YEARS OF EXPERIENCE

Burgundy's reputation for winemaking dates as far back as 51 B.C.

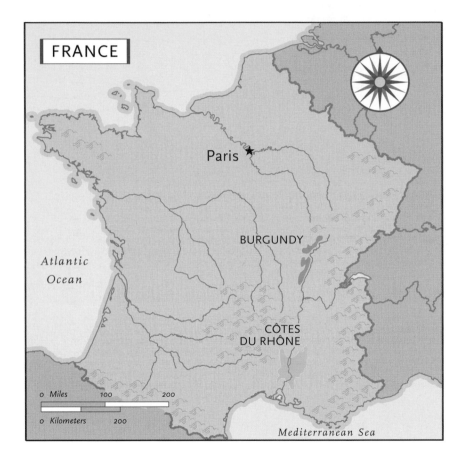

What's so different about red wines (beyond their color)?

We're beginning to see more components in the wines—more complexities. In the white wines, we were looking mainly for the acid/fruit balance, but now, in addition, we're looking for other characteristics, such as tannin.

WHAT'S TANNIN?

Tannin is a natural preservative and is one of the many components that give wine its longevity. It comes from the skins, pits, and stems of the grapes. Another source of tannin is wood, such as the French oak barrels in which some wines are aged or fermented.

A word used to describe the taste sensation of tannin is "astringent." But tannin is not a taste. It's a tactile sensation.

Tannin is also found in strong tea. And what can you add to the tea to make it less astringent? Milk—the fat and the proteins in milk soften the tannin. And so it is with a highly tannic wine. If you take another milk by-product, such as cheese, and have it with wine, it softens the tannin and makes the wine more appealing. Enjoy a beef entrée or one served with a cream sauce and a good bottle of red wine to experience it for yourself.

Why is Burgundy so difficult to understand?

Before we go any further, I must tell you that there are no shortcuts. Burgundy is one of the most difficult subjects in the study of wines. People get confused about Burgundy. They say, "There's so much to know," and "It looks so hard." Yes, there are many vineyards and villages, and they're all important. But there are really only fifteen to twenty-five names you should know if you'd like to understand and speak about Burgundy wines intelligently. Not to worry. I'm going to help you decode all the mysteries of Burgundy: names, regions, and labels.

BLAME NAPOLEON

If you're having trouble understanding the wines of Burgundy, you're not alone. After the French Revolution in 1789, all the vineyards were sold off in small parcels. The Napoleonic Code also called for a law of equal inheritance for the children—continuing to fragment the vineyards even further.

LAND OF 1,000 NAMES

There are more than 1,000 names and more than 110 appellations you must memorize to become a Burgundy wine expert.

What are the main red wine–producing areas of Burgundy?

CÔTE D'OR { CÔTE DE NUITS   BEAUJOLAIS
              CÔTE DE BEAUNE   CÔTE CHÂLONNAISE

THE CÔTE D'OR is only 30 miles long and half a mile wide.

What major grape varieties are used in red Burgundy wines?

The two major grape varieties are Pinot Noir and Gamay. Under Appellation d'Origine Contrôlée laws, all red Burgundies are made from the Pinot Noir grape, except Beaujolais, which is produced from the Gamay grape.

IN AN average year, some 19 million cases of wine are produced in Burgundy. Twelve million cases of that are Beaujolais!

BEAUJOLAIS

- Made from 100 percent Gamay grapes.

- This wine's style is typically light and fruity. It's meant to be consumed young. Beaujolais can be chilled.

- Beaujolais is the best-selling Burgundy in the United States by far, probably because there is so much of it, it's so easy to drink, and it's very affordable. Most bottles cost between eight and twenty dollars, although the price varies with the quality level.

What are the quality levels of Beaujolais?

There are three different quality levels of Beaujolais:

Beaujolais: This basic Beaujolais accounts for the majority of all Beaujolais produced. (Cost: $)

Beaujolais-Villages: This comes from certain villages in Beaujolais. There are thirty-five villages that consistently produce better wines. Most Beaujolais-Villages is a blend of wines from these villages, and usually no particular village name is included on the label. (Cost: $$)

Cru: A cru is actually named for the village that produces this highest quality of Beaujolais. (Cost: $$$$)

There are ten crus (villages):

BROUILLY	JULIÉNAS
CHÉNAS	MORGON
CHIROUBLES	MOULIN-À-VENT
CÔTE DE BROUILLY	RÉGNIÉ
FLEURIE	SAINT-AMOUR

What's Beaujolais Nouveau?

Beaujolais Nouveau is even lighter and fruitier in style than your basic Beaujolais and it is best to drink it young. Isn't that true of all Beaujolais wines? Yes, but Nouveau is different. This "new" Beaujolais is picked, fermented, bottled, and available at your local retailer in a matter of weeks. (I don't know what you call that in your business, but I call it good cash flow in mine. It gives the winemaker a virtually instant return.)

There's another purpose behind Beaujolais Nouveau: Like a preview of a

ALL GRAPES in the Beaujolais region are picked by hand.

movie, it offers the wine-consuming public a sample of the quality of the vintage and style that the winemaker will produce in his regular Beaujolais for release the following spring.

Beaujolais Nouveau is meant to be consumed within six months of bottling. So if you're holding a 2000 Beaujolais Nouveau, now is the time to give it to your "friends."

How long should I keep a Beaujolais?

It depends on the level of quality and the vintage. Beaujolais and Beaujolais-Villages are meant to last between one and three years. Crus can last longer because they are more complex: they have more fruit and tannin. I've tasted Beaujolais crus that were more than ten years old and still in excellent condition. This is the exception, though, not the rule.

Which shippers/producers should I look for when buying Beaujolais?

BOUCHARD   DROUHIN   DUBOEUF

JADOT   MOMMESSIN

BEST BETS FOR RECENT VINTAGES OF BEAUJOLAIS

2002*   2003*   2005*   2006

Note: * signifies exceptional vintage

ONE-THIRD of the Beaujolais grapes are used to make Beaujolais Nouveau.

"Beaujolais is one of the very few red wines that can be drunk as a white. Beaujolais is my daily drink. And sometimes I blend one-half water to the wine. It is the most refreshing drink in the world."

—DIDIER MOMMESSIN

TO CHILL OR NOT TO CHILL?

As a young student studying wines in Burgundy, I visited the Beaujolais region, excited and naive. In one of the villages, I stopped at a bistro and ordered a glass of Beaujolais. (A good choice on my part, don't you think?) The waiter brought the glass of Beaujolais and it was chilled, and I thought that these people had not read the right books! Every wine book I'd ever read always said you serve red wines at room temperature and white wines chilled.

Obviously, I learned from my experience that when it comes to Beaujolais Nouveau, Beaujolais, and Beaujolais-Villages, it's a good idea to give them a slight chill to bring out the fruit and liveliness (acidity) of the wines. That is why Beaujolais is my favorite red wine to have during the summer.

However, to my taste, the Beaujolais crus have more fruit and more tannin and are best served at room temperature.

CÔTE CHÂLONNAISE

Now we're getting into classic Pinot Noir wines that offer tremendous value.

You should know three villages from this area:

Mercurey: (95 percent red)

Givry: (90 percent red)

Rully: (50 percent red)

Mercurey is the most important, producing wines of high quality. Because they are not well known in the United States, Mercurey wines are often a very good buy.

Which shippers/producers should I look for when buying wines from the Côte Châlonnaise?

MERCUREY	FAIVELEY
	DOMAINE DE SUREMAIN
	MICHEL JUILLOT
GIVRY	DOMAINE THENARD
	DOMAINE JABLOT
	LOUIS LATOUR
RULLY	ANTONIN RODET

CÔTE D'OR

Here is the heart of Burgundy. The Côte d'Or (pronounced "coat door") means "golden slope." This region gets its name from the color of the foliage on the hillside, which in autumn is literally golden, as well as the income it brings to the winemakers. The area is very small and its best wines are among the priciest in the world. If you are looking for a $7.99 everyday bottle of wine, this is not the place that will produce it.

What's the best way to understand the wines of the Côte d'Or?

First, you need to know that these wines are distinguished by quality levels—generic, Village, Premier Cru vineyards, and Grand Cru vineyards. Let's look at the quality levels with the double pyramid shown below. As you can see, not much Grand Cru wine is produced, but it is the highest quality and extremely expensive. Generic, on the other hand, is more readily available. Although much is produced, very few generic wines can be classified as outstanding.

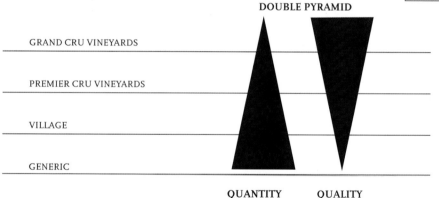

The Côte d'Or is divided into two regions:

Côte de Beaune: Red and white wines (70% red and 30% white).

Côte de Nuits: 95% red wines, the highest-quality red Burgundy wines come from this region.

Another way to understand the wines of the Côte d'Or is to become familiar with the most important villages, Grand Cru vineyards, and some of the Premier Cru vineyards.

GENERIC WINES are labeled simply "Burgundy" or "Bourgogne." A higher level of generic wines will be labeled Côte de Beaune Villages or Côte de Nuits Villages, being a blend of different village wines.

THERE ARE 32 Grand Cru Vineyards:
8 white
24 red
24 from the Côte de Nuits
8 from the Côte de Beaune

Monthélie
Savigny les Beaune
Pernand-Vergelesses
Marsannay
Santenay

THE IMPORTANCE OF SOIL TO BURGUNDY WINES

If you talk to any producers of Burgundy wines, they'll tell you the most important element in making their quality wines is the soil in which the grapes are grown. This, together with the slope of the land and the climatic conditions, determines whether the wine is a Village wine, a Premier Cru, or a Grand Cru. This concept of soil, slope, and climatic conditions in French is known as terroir.

During one of my trips to Burgundy, it rained for five straight days. On the sixth day I saw the workers at the bottom of the slopes with their pails and shovels, collecting the soil that had run down the hillside and returning it to the vineyard. This goes to show the importance of the soil to Burgundy wines.

Côte de Beaune—Red

BY FAR the most-produced red Grand Cru is Corton, representing about 25% of all Grand Cru red wines.

THERE ARE more than 400 Premier Cru vineyards in Burgundy.

MOST IMPORTANT VILLAGES	MY FAVORITE PREMIER CRU VINEYARDS	GRAND CRU VINEYARDS
Aloxe-Corton	Fournières	Corton
	Chaillots	Corton Clos du Roi
		Corton Bressandes
		Corton Renardes
		Corton Maréchaude
Beaune	Grèves	None
	Fèves	
	Marconnets	
	Bressandes	
	Clos des Mouches	
Pommard	Épenots	None
	Rugiens	
Volnay	Caillerets	None
	Santenots	
	Clos des Chênes	
	Taillepieds	

POMMARD
APPELLATION POMMARD CONTRÔLÉE

PRODUCED AND BOTTLED BY
BOUCHARD PÈRE & FILS
ALC 13% BY VOL   CHÂTEAU DE BEAUNE, CÔTE-D'OR, FRANCE   750 ML
PRODUIT DE FRANCE · PRODUCT OF FRANCE
RED BURGUNDY WINE

Côte de Nuits—Red

Finally, we reach the Côte de Nuits. If you're going to spend any time studying your geography, do it now. The majority of Grand Cru vineyards are located in this area.

THE FOUR smallest French appellations are located in the Côte de Nuits:

La Romanée (0.85 acres acres)

La Grande Rue (1.65 acres)

La Romanée Conti (1.8 acres)

Griotte Chambertin (2.75 acres)

MOST IMPORTANT VILLAGES	MY FAVORITE PREMIER CRU VINEYARDS	GRAND CRU VINEYARDS
Gevrey-Chambertin	Clos St-Jacques	Chambertin
	Les Cazetiers	Chambertin Clos de Bèze
	Aux Combottes	Latricières-Chambertin
		Mazis-Chambertin
		Mazoyères-Chambertin
		Ruchottes-Chambertin
		Chapelle-Chambertin
		Charmes-Chambertin
		Griotte-Chambertin
Morey-St-Denis	Ruchots	Clos des Lambrays
	Les Genevrières	Clos de Tart
	Clos des Ormes	Clos St-Denis
		Clos de la Roche
		Bonnes Mares (partial)
Chambolle-Musigny	Les Amoureuses	Musigny
	Charmes	Bonnes Mares (partial)
Vougeot		Clos de Vougeot
Flagey-Échézeaux		Échézeaux
		Grands-Échézeaux
Vosne-Romanée	Beaux-Monts	La Grande-Rue
		Malconsorts
		Romanée-Conti
		La Romanée
		La Tâche
		Richebourg
		Romanée-St-Vivant
Nuits-St-Georges	Les St-Georges	None
	Vaucrains	
	Porets	

CHAMBERTIN CLOS de Bèze was the favorite wine of Napoleon, who is reported to have said: "Nothing makes the future look so rosy as to contemplate it through a glass of Chambertin." Obviously, he ran out of Chambertin at Waterloo!

Joseph Drouhin
RÉCOLTE DU DOMAINE
PREMIER CRU
CHAMBOLLE-MUSIGNY
AMOUREUSES
APPELLATION CONTROLÉE
MIS EN BOUTEILLE PAR JOSEPH DROUHIN NÉGOCIANT ÉLEVEUR À BEAUNE, CÔTE · D'OR, FRANCE, AUX CELLIERS DES ROIS DE FRANCE ET DES DUCS DE BOURGOGNE
13% vol.   FRANCE   75 cl

THE NEWEST GRAND CRU

La Grande-Rue, a vineyard tucked between the Grands Crus La Tâche and Romanée-Conti in Vosne-Romanée, was itself elevated to the Grand Cru level, bringing the number of Grands Crus in the Côte d'Or to 32.

CONTAINS SULFITES   Mise en bouteille au Domaine   PRODUCT OF FRANCE
Nuits-St-Georges
APPELLATION CONTROLÉE
ALC. 12.5% BY VOL.   RED BURGUNDY WINE   750 ML
Domaine Henri Gouges
Nuits-Saint-Georges, Côte-d'Or, France

HAS THIS ever happened to you? In a restaurant, you order a village wine—Gevrey Chambertin, for example—and by mistake the waiter brings you a Grand Cru Chambertin. What would you do?

$
Village Only = Village Wine

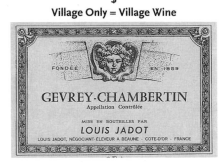

$$
Village + Vineyard
(Clos Saint-Jacques) = Premier Cru

$$$$
Vineyard Only
(Le Chambertin) = Grand Cru

Why are we bothering with all this geography? Must we learn the names of all the villages and vineyards?

I thought you'd never ask. First of all, the geography is important because it helps make you a smart buyer. If you're familiar with the most important villages and vineyards, you're more likely to make an educated purchase.

You really don't have to memorize all the villages and vineyards. I'll let you in on a little secret of how to choose a Burgundy wine and tell at a glance if it's a Village wine, a Premier Cru, or a Grand Cru—usually the label will tip you off in the manner illustrated here:

This is the method I use to teach Burgundy wine. Ask yourself the following:

Where is the wine from? France

What type of wine is it? Burgundy

Which region is it from? Côte d'Or

Which area? Côte de Nuits

Which village is the wine from?
Chambolle-Musigny

Does the label give more details?
Yes, it tells you that the wine is from a vineyard called Musigny, which is one of the thirty-two Grand Cru vineyards.

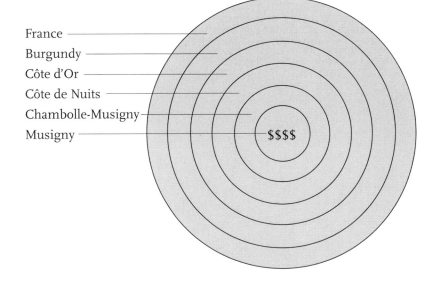

France

Burgundy

Côte d'Or

Côte de Nuits

Chambolle-Musigny

Musigny

$$$$

Has the style of Burgundy wines changed in the last twenty years?

There is a good deal of debate about this in the wine industry. I would have to answer, Yes, it has indeed changed. Winemakers used to make Burgundies to last longer. In fact, you couldn't drink a Burgundy for several years if you wanted to get the fullest flavor. It simply wasn't ready. Today the winemakers of Burgundy are complying with consumer demand for Burgundy they can drink earlier. In America, it seems no one has the patience to wait. A compromise had to be made, however, and that is in the body. Many wines are lighter in style and they can be consumed just a few years after the vintage.

Why are the well-known great Burgundies so expensive?

The answer is simple: supply and demand. The Burgundy growers and shippers of the Côte d'Or have a problem all business people would envy—not enough supply to meet the demand. It has been this way for years and it will continue, because Burgundy is a small region that produces a limited amount of wine. The Bordeaux wine region produces three times as much wine as Burgundy does.

IN THE 1960s, Burgundy wines were fermented and vatted for up to three weeks. Today's Burgundy wines are usually fermented and vatted for six to twelve days.

SLEEPY HOLLOW

There is a village in Westchester County, New York, called Sleepy Hollow. It was originally called North Tarrytown, but the locals, trying to capitalize on "The Legend of Sleepy Hollow" for tourism, changed its name.

And so it is with some villages in Burgundy, as they have added the name of their most famous vineyard to the name of the village. For example, the village of Gevrey became Gevrey-Chambertin, and the village of Puligny became Puligny-Montrachet. Can you figure out the others?

IT'S THE LAW!

Beginning with the 1990 vintage, all Grand Cru Burgundies must include the words "Grand Cru" on the label.

IF YOU DON'T want to be disappointed by the Burgundy wine you select, make sure you know your vintages. Also, due to the delicacy of the Pinot Noir grape, red Burgundies require proper storage, so make sure you buy from a merchant who handles Burgundy wines with care.

ROBERT DROUHIN states that the 1999 vintage for red Burgundy "is one of the greatest vintages I've ever known."

BURGUNDY WINE HARVEST

(average number of cases over a five-year period for red and white)

Regional Appellations	2,136,674
Beaujolais	11,503,617
Côte Châlonnaise	357,539
Côte d'Or (Côte de Nuits)	511,594
Côte d'Or (Côte de Beaune)	1,391,168
Chablis	755,188
Mâconnais	2,136,674
Other Appellations	339,710
Total Burgundy Harvest	19,132,164 cases

BEST BETS FOR RECENT VINTAGES OF CÔTE D'OR

1996*   1997*   1999*   2000
2002*   2003*   2005*

*Note: * signifies exceptional vintage*

HOW IMPORTANT is the producer? Clos de Vougeot is the single largest Grand Cru vineyard in Burgundy, totaling 124 acres, with more than 70 different owners. Each owner makes his own winemaking decisions, such as when to pick the grapes, the style of fermentation, and how long to age the wine in oak. Obviously, all Clos de Vougeot is not created equal.

Who are the most important shippers to look for when buying red Burgundy wine?

BOUCHARD PÈRE ET FILS	LOUIS JADOT
JOSEPH DROUHIN	LOUIS LATOUR
JAFFELIN	LABOURÉ-ROI
CHANSON	

Although 80 percent of Burgundy wine is sold through shippers, some fine estate-bottled wines are available in limited quantities in the United States. Look for the following

DOMAINE CLERGET	DOMAINE LOUIS TRAPET
DOMAINE COMTE DE VOGÜE	DOMAINE MONGEARD-MUGNERET
DOMAINE DANIEL RION	
DOMAINE DE LA ROMANÉE-CONTI	DOMAINE PARENT
DOMAINE DUJAC	DOMAINE PIERRE DAMOY
DOMAINE GEORGES ROUMIER	DOMAINE POTEL
DOMAINE GROFFIER	DOMAINE POUSSE D'OR
DOMAINE HENRI GOUGES	DOMAINE PRINCE DE MÉRODE
DOMAINE HENRI LAMARCHE	DOMAINE TOLLOT-BEAUT
DOMAINE JAYER	DOMAINE VINCENT GIRARDIN
DOMAINE JEAN GRIVOT	MAISON FAIVELEY
DOMAINE LEROY	

FOR FURTHER READING

I recommend *Côte d'Or* by Clive Coates; *Burgundy* by Anthony Hanson; *Making Sense of Burgundy* by Matt Kramer; *The Great Domaines of Burgundy* by Remington Norman; and *Burgundy* by Robert M. Parker Jr.

WINE AND FOOD

To get the most flavor from both the wine and the food, some of Burgundy's famous winemakers offer these suggestions:

ROBERT DROUHIN: *"In my opinion, white wine is never a good accompaniment to red meat, but a light red Burgundy can match a fish course (not shellfish). Otherwise, for light red Burgundies, white meat—not too many spices; partridge, pheasant, and rabbit. For heavier-style wines, lamb and steak are good choices."* Personally, Mr.

Drouhin does not enjoy red Burgundies with cheese—especially goat cheese.

PIERRE HENRY GAGEY (Louis Jadot): *"With red Beaujolais wines, such as Moulin-à-Vent Château des Jacques, for example, a piece of pork like an andouillette from Fleury is beautiful. A Gamay, more fruity and fleshy than Pinot Noir, goes perfectly with this typical meal from our terroir. My favorite food combination with a red Burgundy wine is poulet de bresse demi d'oeil. The very thin flesh of this truffle-filled chicken and the elegance and delicacy from the great Pinot Noir, which come from the best terroir, go together beautifully."*

LOUIS LATOUR: *"With Château Corton Grancey, filet of duck in a red wine sauce. Otherwise, Pinot Noir is good with roast chicken, venison, and beef. Mature wines are a perfect combination for our local cheeses, Chambertin and Citeaux."*

The Red Wines of the Rhône Valley

As a former sommelier, I was often asked to recommend a big, robust red Burgundy wine to complement a rack of lamb or filet mignon. To the customers' surprise, I didn't recommend a Burgundy at all. The best bet is a Rhône wine, which is typically a bigger and fuller wine than one from Burgundy, and usually has a higher alcoholic content. The reason for these characteristics is quite simple. It all goes back to location and geography.

Where's the Rhône Valley?

The Rhône Valley is in southeastern France, south of the Burgundy region, where the climate is hot and the conditions are sunny. The extra sun gives the grapes more sugar, which, as we have discussed, boosts the level of alcohol. The soil is full of rocks that retain the intense summer heat during both day and night.

Winemakers of the Rhône Valley are required by law to make sure their wines have a specified amount of alcohol. For example, the minimum alcoholic content required by the AOC is 10.5 percent for Côtes du Rhône and 13.5 percent for Châteauneuf-du-Pape.

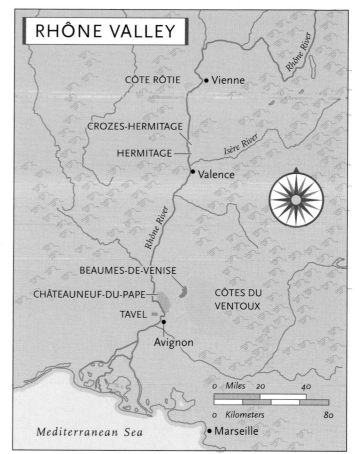

What are the different quality levels of the Rhône Valley?

	% of Production
1. Côtes du Rhône ($)	58%
2. Côtes du Rhône Villages ($$)	8%
3. Côtes du Rhône Crus (specific regions) ($$$$)	10%
Northern Rhône	
Southern Rhône	
4. Other appellations	24%

OF ALL the wines made in the Rhône Valley, 91% are red, 6% are rosé, and 3% are white.

SOME OF the oldest vineyards in France are in the Rhône Valley. Hermitage, for example, has been in existence for more than 2,000 years.

CÔTES DU RHÔNE wine can be produced from grapes grown in either or both the northern and southern Rhône regions. More than 90% of all Côtes du Rhône wines come from the southern region.

TWO OTHER important red grapes in the Rhône Valley are:

Cinsault

Mourvèdre

THE 13 CRUS OF THE RHÔNE VALLEY

North

Château-Grillet (white)

Condrieu (white)

Cornas

Côte-Rôtie

Crozes-Hermitage

Hermitage

St-Joseph

St-Peray

South

Châteauneuf-du-Pape

Gigondas

Lirac

Tavel (rosé)

Vacqueyras

What are the winemaking regions in the Rhône Valley?

The region is divided into two distinct areas: northern and southern Rhône. The most famous red wines that come from the northern region are:

CROZES-HERMITAGE (3,000+ ACRES)
CÔTE RÔTIE (555 ACRES)
HERMITAGE (324 ACRES)

The most famous red wines from the southern region are **Châteauneuf-du-Pape (7,822 acres)** and **Gigondas (3,036 acres)**; one of the best French rosés is called **Tavel**.

Two distinct microclimates distinguish the north from the south. It is important for you to understand that these areas make distinctly different wines because of:

- Soil
- Location
- Different grape varieties used in making the wines of each area.

What are the main red-grape varieties grown in the Rhône Valley?

The two major grape varieties in the Rhône Valley are:

GRENACHE
SYRAH

Which wines are made from these grapes?

The Côte Rôtie, Hermitage, and Crozes-Hermitage from the north are made primarily from the Syrah grape. These are the biggest and fullest wines from that region.

For Châteauneuf-du-Pape, as many as thirteen different grape varieties may be included in the blend. But the best producers use a greater percentage of Grenache and Syrah in the blend.

What's Tavel?

It's a rosé—an unusually dry rosé, which distinguishes it from most others. It's made primarily from the Grenache grape, although nine grape varieties can be used in the blend. When you come right down to it, Tavel is just like a red wine, with all the red-wine components but less color. How do they make a rosé wine with red-wine characteristics but less color? It's all in the vatting process.

What's the difference between "short-vatted" and "long-vatted" wines?

When a wine is "short-vatted," the skins are allowed to ferment with the must (grape juice) for a short period of time—only long enough to impart that rosé color. It's just the opposite when a winemaker is producing red Rhône wines, such as Châteauneuf-du-Pape of Hemitage. The grape skins are allowed to ferment longer with the must, giving a rich, ruby color to the wine.

What's the difference between a $25 bottle of Châteauneuf-du-Pape and a $75 bottle of Châteauneuf-du-Pape?

A winemaker is permitted to use thirteen different grapes for his Châteauneuf-du-Pape recipe, as I mentioned earlier. It's only logical, then, that the winemaker who uses a lot of the best grapes (which is equivalent to cooking with the finest ingredients) will produce the best-tasting—and the most expensive—wine.

For example, a $25 bottle of Châteauneuf-du-Pape may contain only 20 percent of top-quality grapes (Grenache, Mourvèdre, Syrah, and Cinsault) and 80 percent of lesser-quality grapes; a $75 bottle may contain 90 percent of the top-quality grapes and 10 percent of others.

BEST BETS FOR RED RHÔNE VALLEY WINES

North: 1995   1996   1997   1998   1999*   2000
2001   2003*   2004   2005   2006*   2007*
South: 1995   1998*   1999   2000*   2001*
2003*   2004*   2005*   2006*   2007*

Note: * signifies exceptional vintage

How do I buy a red Rhône wine?

You should first decide if you prefer a light Côtes du Rhône wine or a bigger, more flavorful one, such as an Hermitage. Then you must consider the vintage and the producer. Two of the oldest and best-known firms are M. Chapoutier and Paul Jaboulet Aîné. Also look for the producers Guigal, Chave, Beaucastel, Domaine du Vieux Télégraphe, Alain Graillot, Clos des Papes, Roger Sabon & Fils, Mont Redon, Le Vieux Donjon, Domaine du Pégau, and Château Rayas.

RHÔNE VALLEY vintages can be tricky: A good year in the north may be a bad year in the south, and vice versa.

OLDER GREAT VINTAGES

North: 1983, 1985, 1988, 1989, 1990, 1991
South: 1985, 1988, 1989, 1990

A SIMPLE Côtes du Rhône is similar to a Beaujolais—except the Côtes du Rhône has more body and alcohol. A Beaujolais, by AOC standards, must contain a minimum of 9% alcohol; a Côtes du Rhône, 10.5%.

Grenache	Muscardin
Syrah	Vaccarèse
Mourvèdre	Picardin
Cinsault	Picardin
Picpoul	Clairette
Terret	Roussanne
Counoise	Bourboulenc

The first four represent 92% of the grapes used, with Grenache by far the most.

THERE IS no official classification for Rhône Valley wines.

SUNSHINE QUOTIENT

	(hours per year)
Burgundy	2,000
Bordeaux	2,050
Châteauneuf-du-Pape	2,750

THE PAPAL coat of arms from medieval times appears on some Châteauneuf-du-Pape bottles. Only owners of vineyards are permitted to use this coat of arms on the label.

When should I drink my Rhône wine?

Tavel: Within two years.

Côtes du Rhône: Within three years.

Crozes-Hermitage: Within five years.

Châteauneuf-du-Pape: After five years, but higher-quality Châteauneuf-du-Pape is better after ten years.

Hermitage: Seven to eight years, but best after fifteen, in a great year.

RHÔNE VALLEY HARVEST

(average number of cases over a five-year period)

Côtes du Rhône regional appellation	18.8 million
Northern and southern Crus	2.6 million
Côtes du Rhône-Villages appellation	1.6 million
Total Rhône Valley harvest	23.0 million cases

MIS EN BOUTEILLE DU CHATEAU

Château de Beaucastel

CHATEAUNEUF-DU-PAPE
APPELLATION CHATEAUNEUF-DU-PAPE CONTROLÉE

Sté FERMIÈRE DES VIGNOBLES PIERRE PERRIN
AU CHATEAU DE BEAUCASTEL COURTHEZON (Vse) FRANCE 750 ml
ALC. 13.5% BY VOL.
PRODUCE OF FRANCE
IMPORTED BY **Vineyard Brands, Inc.** BIRMINGHAM, AL
SHIPPED BY ROBERT HAAS SELECTIONS, FRANCE

THE RED RHÔNE VALLEY ROUNDUP

NORTHERN WINES	**SOUTHERN WINES**
Côte Rôtie	*Châteauneuf-du-Pape*
Hermitage	*Tavel*
Crozes-Hermitage	*Côtes du Rhône*
St. Joseph	*Côtes du Rhône-Villages*
Cornas	*Côtes du Ventoux*
	Gigondas

GRAPE	**MAJOR GRAPES**
Syrah	*Grenache*
	Syrah
	Cinsault
	Mourvèdre

WINE AND FOOD

Jean Pierre and François Perrin (*Château de Beaucastel and La Vieille Ferme*): *"The white wines of Château de Beaucastel can be drunk either very young—in the first three or four years—or should be kept for ten years or more. The combination of white meat with truffles and mushrooms is an exquisite possibility.*

"Red Rhône wines achieve their perfection from ten years and beyond, and are best when combined with game and other meats with a strong flavor.

"A good dinner could be wild mushroom soup and truffles with a white Beaucastel and stew of wild hare á la royale (with foie gras and truffles) served with a red Château de Beaucastel."

Michel Chapoutier: *With a Côtes du Rhône wine, he recommends poultry, light meats, and cheese. Côte Rôtie goes well with white meats and small game. Châteauneuf-du-Pape complements the ripest of cheese, the richest venison, and the most lavish civet of wild boar. An Hermitage is suitable with beef, game, and any full-flavored cheese. Tavel rosé is excellent with white meat and poultry.*

Frédéric Jaboulet: *"My granddad drinks a bottle of Côtes du Rhône a day and he's in his eighties. It's good for youth. It goes with everything except old fish,"* he jokes.

More specifically, Mr. Jaboulet says, "Hermitage is good with wild boar and mushrooms. A Crozes Hermitage, particularly our Domaine de Thalabert, complements venison or roast rabbit in a cream sauce, but you have to be very careful with the sauce and the weight of the wine.

"Beef ribs and rice go well with a Côtes du Rhône, as does a game bird like roast quail. Tavel, slightly chilled, is refreshing with a summer salad. Muscat de Beaumes-de-Venise, of course, is a beautiful match with foie gras."

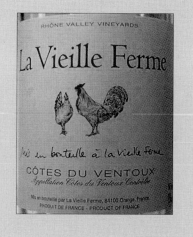

Hermitage is the best and the longest-lived of the Rhône wines. In a great vintage, Hermitage wines can last for fifty years.

U.S. imports of Rhône wines have risen more than 200% in the last five years.

The two most famous white wines of the Rhône Valley are called Condrieu and Château Grillet. Both are made from the grape variety called Viognier.

There is a white Châteauneuf-du-Pape and a white Hermitage, but only a few thousand cases are produced each year.

Châteauneuf-du-Pape means "new castle of the Pope," so named for the palace in the Rhône city of Avignon in which Pope Clément V (the first French pope) resided in the 14th century.

A good value is Côtes du Ventoux. One of the most widely available wines to look for in this category is La Vieille Ferme. For those who prefer sweet wines, try Beaumes-de-Venise, made from the Muscat grape.

Questions for Class Four: The Red Wines of Burgundy and the Rhône Valley

    a. Gamay   ___ Bordeaux

    b. Pinot Noir   ___ Tuscany

    c. Tempranillo   ___ Rhône

    d. Sangiovese   ___ Piedmont

    e. Merlot   ___ Rioja

    f. Cabernet Sauvignon   ___ Beaujolais

    g. Nebbiolo   ___ Champagne

    h. Syrah   ___ Burgundy

The Red Wines of Bordeaux

APPELLATIONS • GRAPE VARIETIES • QUALITY LEVELS •

THE GREAT RED WINES OF BORDEAUX • MÉDOC • GRAVES •

POMEROL • ST-ÉMILION • CHOOSING A RED BORDEAUX

The Red Wines of Bordeaux

THIS PROVINCE OF FRANCE is rich with excitement and history, and the best part is that the wines speak for themselves. You'll find this region much easier to learn about than Burgundy. For one thing, the plots of land are bigger, and they're owned by fewer landholders. And, as Samuel Johnson once said, "He who aspires to be a serious wine drinker must drink claret." That is, the dry red table wine known as Bordeaux.

Some fifty-seven wine regions in Bordeaux produce high-quality wines that are allowed to carry the AOC designation on the label. Of these fifty-seven places, four stand out in my mind for red wine:

> **Médoc:** 40,199 acres (produces only red wines)
>
> **Pomerol:** 1,846 acres (produces only red wines)

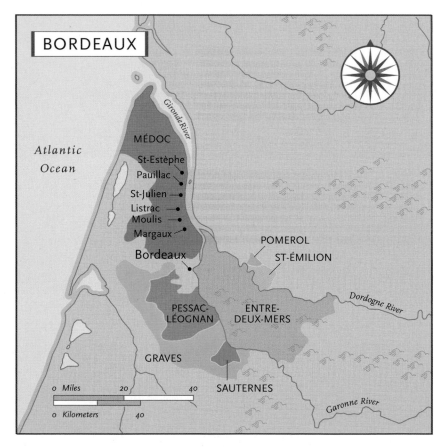

THE ENGLISH word *claret* refers to dry red wines from Bordeaux.

BORDEAUX IS much larger in acreage than Burgundy.

OF ALL the AOC wines of France, 27% come from the Bordeaux region.

IN DOLLAR value, the United States is the fourth-largest importer of Bordeaux wines.

TODAY, 84% of Bordeaux wine is red and 16% is white.

UNTIL 1970, BORDEAUX regularly produced more white wine than red.

Graves/Pessac-Léognan: 9,855 acres (produces both red and dry white wines)

St-Émilion: 23,384 acres (produces only red wines)

In the Médoc, there are seven important inner appellations you should be familiar with:

Haut Médoc   St-Estèphe   Pauillac   St-Julien
Margaux   Moulis   Listrac

Which grape varieties are grown in Bordeaux?

The three major grapes are:

Merlot

Cabernet Sauvignon

Cabernet Franc

Unlike Burgundy, where the winemaker must use 100 percent Pinot Noir to make most red wines (100 percent Gamay for Beaujolais), in Bordeaux the red wines are almost always made from a blend of grapes.

What are the different quality levels of Bordeaux wine?

Bordeaux ($): This is the lowest level of AOC wine in Bordeaux—wines that are nice, inexpensive, and consistent "drinking" wines. These are sometimes known as "proprietary" wines—wines known by what you could almost call a brand name, such as Mouton-Cadet, rather than by the particular region or vineyard. These are usually the least expensive AOC wines in Bordeaux.

Region ($$): Regional wines come from one of the fifty-seven different regions. Only grapes and wines made in those areas can be called by their regional names: Pauillac and St-Émilion, for example. These wines are more expensive than those labeled simply Bordeaux.

Region + Château ($$–$$$$): Château wines are the products of individual vineyards. There are more than 7,000 châteaux in Bordeaux. As far back as 1855, Bordeaux officially classified the quality levels of some of its châteaux. Hundreds have been officially recognized for their quality. In the Médoc, for example, the 61 highest-level châteaux are called Grand Cru Classé. There are also 247 châteaux in the Médoc that are entitled to be called Cru Bourgeois, a step below Grand Cru Classé. Other areas, such as St-Émilion and Graves, have their own classification systems.

THE GRAVES region produces 60% red wine, 40% white wine.

IN 1987, a communal appellation was established to create a higher-level appellation in the northern Graves region. It's called Pessac-Léognan (for both reds and whites).

TAKE A LOOK at the map on page 118. As a general rule of thumb, red wines from the villages and regions on the left bank of the rivers primarily use the Cabernet Sauvignon grape and on the right bank they use Merlot.

IN ALL of Bordeaux, there are some 99,000 acres of Merlot, 62,000 acres of Cabernet Sauvignon, and 32,000 acres of Cabernet Franc.

TWO OTHER grapes that are sometimes used in the blending of Bordeaux wines are Petit Verdot and Malbec.

PROPRIETARY WINES you may be familiar with:

Lauretan	Baron Philippe
Lacour Pavillon	Michel Lynch
Mouton-Cadet	

THE MAJOR shippers of regional wines from Bordeaux are:

Barton & Guestier (B & G)
Cordier
Dourthe Kressmann
Eschenauer
Sichel
Yvon Mau
Ets J-P Moueix
Baron Philippe de Rothschild
Borie-Manoux
Dulong

WINES PRICED between $8 and $25 represent 80% of the total production of Bordeaux.

THE TYPICAL Médoc blend can vary from 60% to 80% Cabernet Sauvignon, 25% to 40% Merlot, and 10% to 20% Cabernet Franc.

IF YOU see a château on the label, French law dictates that the château really exists and is the château of that wine-maker. What you see is what you get.

BORDEAUX (PROPRIETARY)
APPELLATION BORDEAUX
CONTRÔLLÉE

REGIONAL
APPELLATION PAUILLAC
CONTRÔLLÉE

CHÂTEAU
APPELLATION PAUILLAC
CONTRÔLLÉE WITH
CHÂTEAU NAME

All three of these wines are owned by the same family, the Rothschilds, who also own Château Mouton-Rothschild.

ACCORDING TO French law, a château is a house attached to a vineyard having a specific number of acres, as well as having winemaking and storage facilities on the property. A wine may not be called a château wine unless it meets these criteria. The terms *domaine*, *clos*, and *cru* are also used.

Here are the major Bordeaux classifications:

Médoc (Grand Crus Classé): 1855; sixty-one châteaux
Médoc (Crus Bourgeois): 1920, revised 1932, 1978, and 2003; 247 châteaux
Graves (Grand Crus Classé): 1959; sixteen châteaux
Pomerol: No official classification
St-Émilion: 1955, revised 1996, revised 2006; fifteen Premiers Grand Crus Classé and forty-six Grand Crus Classé

What is a château?

When most people think of a château, they picture a grandiose home filled with Persian rugs and valuable antiques and surrounded by rolling hills of vineyards. Well, I'm sorry to shatter your dreams (and the dictionary's definition), but most châteaux are not like that at all. Yes, a château could be a mansion on a large estate, but it could also be a modest home with a two-car garage.

Château wines are usually considered the best-quality wines from Bordeaux. They are the most expensive wines; some examples of the best known of the Grand Cru Classé command the highest wine prices in the world!

Let's take a closer look at the châteaux. One fact I've learned from my years of teaching wine is that no one wants to memorize the names of thousands of châteaux, so I'll shorten the list by starting with the most important classification in Bordeaux.

THE GREAT RED WINES OF BORDEAUX

MÉDOC: GRAND CRU CLASSÉ, 1855; 61 CHÂTEAUX

When and how were the château wines classified?

More than 150 years ago in the Médoc region of Bordeaux, a wine classification was established. Brokers from the wine industry were asked by Napoleon III to select the best wines to represent France in the International Exposition of 1855. The top Médoc wines were ranked according to price, which at that time was directly related to quality. (After all, don't we class everything, from cars to restaurants?) The brokers agreed, provided the classification would never become official. Voilà! Refer to the chart on page 123 for the Official Classification of 1855.

Is the 1855 classification still in use today?

Every wine person knows about the 1855 classification, but much has changed over the last century and a half. Some vineyards have doubled or tripled their production by buying up their neighbors' land, which is permitted by law. Obviously the chateaux have seen many changes of ownership. And, like all businesses, Bordeaux has seen good times and bad times.

A case in point was in the early 1970s, when Bordeaux wines were having a difficult time financially, even at the highest level. At Château Margaux, the well-known first-growth vineyard, the quality of the wine fell off from its traditional excellence for a while when the family that owned the château wasn't putting enough money and time into the vineyard. In 1977, Château Margaux was sold to a Greek-French family (named Mentzelopoulos) for $16 million, and since then the quality of the wine has risen even beyond its first-growth standards.

Château Gloria, in the commune of St-Julien, is an example of a vineyard that didn't exist at the time of the 1855 classification. The late mayor of St-Julien, Henri Martin, bought many parcels of second-growth vineyards. As a result, he produced top-quality wine that is not included in the 1855 classification.

It's also important to consider the techniques used to make wine today. They're a lot different from those used in 1855. Once again, the outcome is better wine. As you can see, some of the châteaux listed in the 1855 classification deserve a lesser ranking, while others deserve a better one.

That said, I believe, even after its 150th anniversary, in most cases it is still a very valid classification in terms of quality and price.

"For a given vintage there is quite a consistent ratio between the prices of the different classes, which is of considerable help to the trade. So a fifth growth would always sell at about half the price of a second. The thirds and fourths would get prices halfway between the seconds and the fifths. The first growths are getting about 25% over the second growths."
—CH. COCKS, Bordeaux et Ses Vins, 1868

THE GRAND CRU Classé châteaux represent less than 5% of the total volume of Bordeaux.

The Official (1855) Classification of the Great Red Wines of Bordeaux

THE MÉDOC

FIRST GROWTHS—PREMIERS CRUS (5)

Vineyard	AOC
Château Lafite-Rothschild	Pauillac
Château Latour	Pauillac
Château Margaux	Margaux
Château Haut-Brion	Pessac-Léognan (Graves)
Château Mouton-Rothschild	Pauillac

SECOND GROWTHS—DEUXIÈMES CRUS (14)

Vineyard	AOC
Château Rausan-Ségla	Margaux
Château Rausan Gassies	Margaux
Château Léoville-Las-Cases	St-Julien
Château Léoville-Poyferré	St-Julien
Château Léoville-Barton	St-Julien
Château Durfort-Vivens	Margaux
Château Lascombes	Margaux
Château Gruaud-Larose	St-Julien
Château Brane-Cantenac	Margaux
Château Pichon-Longueville-Baron	Pauillac
Château Pichon-Longueville-Lalande	Pauillac
Château Ducru-Beaucaillou	St-Julien
Château Cos d'Estournel	St-Estèphe
Château Montrose	St-Estèphe

THIRD GROWTHS—TROISIÈMES CRUS (14)

Vineyard	AOC
Château Giscours	Margaux
Château Kirwan	Margaux
Château d'Issan	Margaux
Château Lagrange	St-Julien
Château Langoa-Barton	St-Julien
Château Malescot-St-Exupéry	Margaux
Château Cantenac-Brown	Margaux
Château Palmer	Margaux
Château La Lagune	Haut-Médoc
Château Desmirail	Margaux
Château Calon-Ségur	St-Estèphe
Château Ferrière	Margaux
Château d'Alesme (formerly Marquis d'Alesme)	Margaux
Château Boyd-Cantenac	Margaux

FOURTH GROWTHS—QUATRIÈMES CRUS (10)

Vineyard	AOC
Château St-Pierre	St-Julien
Château Branaire-Ducru	St-Julien
Château Talbot	St-Julien
Château Duhart-Milon-Rothschild	Pauillac
Château Pouget	Margaux
Château La Tour-Carnet	Haut-Médoc
Château Lafon-Rochet	St-Estèphe
Château Beychevelle	St-Julien
Château Prieuré-Lichine	Margaux
Château Marquis de Terme	Margaux

FIFTH GROWTHS—CINQUIÈMES CRUS (18)

Vineyard	AOC
Château Pontet-Canet	Pauillac
Château Batailley	Pauillac
Château Grand-Puy-Lacoste	Pauillac
Château Grand-Puy-Ducasse	Pauillac
Château Haut-Batailley	Pauillac
Château Lynch-Bages	Pauillac
Château Lynch-Moussas	Pauillac
Château Dauzac	Haut-Médoc
Château d'Armailhac (called Château Mouton-Baron-Philippe from 1956 to 1988)	Pauillac
Château du Tertre	Margaux
Château Haut-Bages-Libéral	Pauillac
Château Pédesclaux	Pauillac
Château Belgrave	Haut-Médoc
Château Camensac	Haut-Médoc
Château Cos Labory	St-Estèphe
Château Clerc-Milon-Rothschild	Pauillac
Château Croizet Bages	Pauillac
Château Cantemerle	Haut-Médoc

ON THE 1945 Mouton-Rothschild bottle there is a big V that stands for "victory" and the end of World War II. Since 1924, Philippe de Rothschild asked a different artist to design his labels, a tradition continued by the Baroness Philippine, his daughter. Some of the most famous artists in the world have agreed to have their work grace the Mouton label, including:

Jean Cocteau—1947
Salvador Dalí—1958
Henry Moore—1964
Joan Miró—1969
Marc Chagall—1970
Pablo Picasso—1973
Robert Motherwell—1974
Andy Warhol—1975
John Huston—1982
Saul Steinberg—1983
Keith Haring—1988
Francis Bacon—1990
Setsuko—1991
Antoni Tàpies—1995
Gu Gan—1996
Robert Wilson—2001
HRH Charles, Prince of Wales—2004

"The classified growths are divided in five classes and the price difference from one class to another is about 12 percent."

—WILLIAM FRANK,
Traité Sur les Vins du Médoc, 1855

HAVE THERE EVER BEEN ANY CHANGES IN THE 1855 CLASSIFICATION?

Yes, but only once, in 1973. Château Mouton-Rothschild was elevated from a second-growth to a first-growth vineyard. There's a little story behind that.

EXCEPTION TO THE RULE . . .

In 1920, when the Baron Philippe de Rothschild took over the family vineyard, he couldn't accept the fact that back in 1855 his château had been rated a second growth. He thought it should have been classed a first growth from the beginning—and he fought to get to the top for some fifty years. While the baron's wine was classified as a second growth, his motto was:

> *First, I cannot be.*
> *Second, I do not deign to be.*
> *Mouton, I am.*

When his wine was elevated to a first growth in 1973, Rothschild replaced the motto with a new one:

> *First, I am.*
> *Second, I was.*
> *But Mouton does not change.*

PABLO PICASSO, 1973

Is there an easier way to understand the 1855 classification?

I've always found the 1855 classification to be a little cumbersome, so one day I sat down and drew up my own chart. I separated the classification into growths (first, second, third, etc.) and then I listed the communes (Pauillac, Margaux, St-Julien, etc.) and set down the number of distinctive vineyards in each one. My chart shows which communes of Bordeaux have the most first growths—all the way down to fifth growths. It also shows which commune corners the market on all growths. Since I was inspired to figure this out during baseball's World Series, I call my chart a box score of the 1855 classification.

KEVIN ZRALY'S BOX SCORE OF THE 1855 CLASSIFICATION

Commune	1st	2nd	3rd	4th	5th	Total
Margaux	I	5	10	3	2	21
Pauillac	3	2	0	1	12	18
St-Julien	0	5	2	4	0	11
St-Estèphe	0	2	1	1	1	5
Haut-Médoc	0	0	1	1	3	5
Graves	1	0	0	0	0	1
Total Châteaux	5	14	14	10	18	61

A quick glance at my box score gives you some instant facts that may guide you when you want to buy a Bordeaux wine from Médoc.

Tallying the score, Pauillac has three of the five first growths and twelve fifth growths. Margaux practically clean-sweeps the third growths. In fact, Margaux is the overall winner, because it has the greatest number of classed vineyards in all of Médoc. Margaux is also the only area to have a château rated in each category. St-Julien has no first or fifth growths, but is very strong in the second and fourth.

LAROSE-TRINTAUDON is the largest vineyard in the Médoc area, making nearly 100,000 cases of wine per year.

MÉDOC CRU BOURGEOIS: 1920, REVISED 1932, 1978, AND 2003; 247 CHÂTEAUX

What does Cru Bourgeois mean?

The Crus Bourgeois of the Médoc are châteaux that were originally classified in 1920, and not in the 1855 classification. In 1932 there were 444 properties listed, but by 1962 there were only 94 members. Today there are 247. The latest classification of Crus Bourgeois of the Médoc and Haut-Médoc was in 2003. Because of the high quality of the 2000, 2003, and 2005 vintages, some of the best values in wine today are in the Cru Bourgeois classification.

The following is a partial list of Crus Bourgeois to look for:

CHÂTEAU D'ANGLUDET	CHÂTEAU PATACHE D'AUX
CHÂTEAU LES ORMES-DE-PEZ	CHÂTEAU LA CARDONNE
CHÂTEAU LES ORMES-SORBET	CHÂTEAU POUJEAUX
CHÂTEAU PHÉLAN-SÉGUR	CHÂTEAU SIRAN
CHÂTEAU COUFRAN	CHÂTEAU DE PEZ
CHÂTEAU CHASSE-SPLEEN	CHÂTEAU PONTENSAC
CHÂTEAU MEYNEY	CHÂTEAU GLORIA
CHÂTEAU SOCIANDO-MALLET	CHÂTEAU PIBRAN
CHÂTEAU FOURCAS-HOSTEN	CHÂTEAU MONBRISON
CHÂTEAU LAROSE-TRINTAUDON	CHÂTEAU VIEUX ROBIN
CHÂTEAU GREYSAC	CHÂTEAU LABÉGORCE-ZÉDÉ
CHÂTEAU MARBUZET	CHÂTEAU DE LAMARQUE
CHÂTEAU HAUT-MARBUZET	

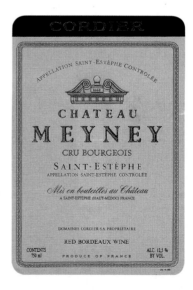

THE RED wines of Pomerol tend to be softer, fruitier, and ready to drink sooner than the Médoc wines.

THE MAJOR grape used to produce wine in the Pomerol region is Merlot. Very little Cabernet Sauvignon is used in these wines.

IT TAKES Château Pétrus one year to make as much wine as Gallo makes in six minutes.

THE VINEYARD at Château Pétrus makes one of the most expensive wines of Bordeaux. It's planted with 95% Merlot.

GRAVES: 1959 GRAND CRUS CLASSÉ; 12 CHÂTEAUX

The most famous château—we have already seen it in the 1855 classification—is Château Haut-Brion. Other good red Graves classified in 1959 as Grands Crus Classés are:

CHÂTEAU BOUSCAUT
CHÂTEAU HAUT-BAILLY
CHÂTEAU CARBONNIEUX
DOMAINE DE CHEVALIER
CHÂTEAU DE FIEUZAL
CHÂTEAU OLIVIER
CHÂTEAU MALARTIC-LAGRAVIÈRE
CHÂTEAU LA TOUR-MARTILLAC
CHÂTEAU SMITH-HAUT-LAFITTE
CHÂTEAU PAPE-CLÉMENT
CHÂTEAU LA MISSION-HAUT-BRION

POMEROL: NO OFFICIAL CLASSIFICATION

This is the smallest of the top red-wine districts in Bordeaux. Pomerol produces only 15 percent as much wine as St-Émilion; as a result, Pomerol wines are relatively scarce. And if you do find them, they'll be expensive. Although no official classification exists, here's a list of some of the finest Pomerols on the market:

CHÂTEAU PÉTRUS
CHÂTEAU LE PIN
CHÂTEAU LA CONSEILLANTE
CHÂTEAU BEAUREGARD
CHÂTEAU PETIT-VILLAGE
CHÂTEAU NÉNIN
CHÂTEAU TROTANOY
CHÂTEAU LATOUR-À-POMEROL
CHÂTEAU L'ÉVANGILE
CHÂTEAU BOURGNEUF
VIEUX CHÂTEAU-CERTAN
CHÂTEAU CLINET
CHÂTEAU LA POINTE
CHÂTEAU L'ÉGLISE CLINET
CHÂTEAU LAFLEUR
CHÂTEAU PLINCE

CHÂTEAU LA FLEUR-PÉTRUS
CHÂTEAU GAZIN

ST-ÉMILION: 1955, REVISED 1996, REVISED 2006; 15 PREMIERS GRANDS CRUS CLASSÉS AND 46 CHÂTEAUX GRANDS CRUS CLASSÉS

This area produces about two-thirds as much wine as the entire Médoc, and St. Emilion is one of the most beautiful villages in France (in my opinion). The wines of St-Émilion were finally classified officially in 1955, one century after the Médoc classification. There are fifteen first growths comparable to the cru classé wines of the Médoc.

THE FIFTEEN FIRST GROWTHS OF ST-ÉMILION
(PREMIERS GRANDS CRUS CLASSÉS)

Château Ausone
Château Cheval Blanc
Château Angélus
Château Beau-Séjour-Bécot
Château Beauséjour-
    Duffau-Lagarrosse
Château Belair
Château Figeac

Château Canon
Château Magdelaine
Château La Gaffelière
Château Troplong Mondot
Château Trottevieille
Château Pavie
Clos Fourtet
Château Pavie Marquin

Important Grands Crus Classés and other St-Émilion wines available in the United States:

Château Canon-La-Gaffelière
Château La Tour-Figeac
Château Trimoulet
Château Dassault
Château Monbousquet
Château Tertre Roteboeuf

Château Trotanoy
Château Faugères
Château Haut-Corbin
Château Grand-Mayne

GRAPE VARIETIES OF ST-ÉMILION

Merlot	70%
Cabernet Franc	25%
Cabernet Sauvignon	5%

SOME OTHER appellations to look for in Bordeaux red wines:
    Fronsac
    Côtes de Blaye
    Côtes de Bourg

THE 1999 VINTAGE was the largest harvest ever in Bordeaux.

ABOUT VINTAGES, Alexis Lichine, a noted wine expert, once said: "Great vintages take time to mature. Lesser wines mature faster than the greater ones. . . . Patience is needed for great vintages, hence the usefulness and enjoyment of lesser vintages." He summed up: "Often vintages which have a poorer rating—if young—will give a greater enjoyment than a better-rated vintage—if young."

DRINK YOUR lighter vintages such as 1999 and 2001 Bordeaux while you wait patiently for your great vintages such as 2000, 2003 and 2005 to mature.

THE EARLIEST harvest since 1893 was 2003.

THE 2003 VINTAGE suffered summer heat waves, fierce storms, and hail, reducing production to the lowest since 1991.

Now that you know all the greatest red-wine regions of Bordeaux, let me take you a step further and show you some of the best vintages.

BORDEAUX VINTAGES

"Left Bank"
Médoc/St-Julien/Margaux/
Pauillac/St-Estèphe/Graves

GOOD VINTAGES	GREAT VINTAGES	OLDER GREAT VINTAGES
1994	1990*	1982*
1997	1995	1985
1998	1996	1986
1999	2000*	1989
2001	2003	
2002	2005*	
2004		
2006		

"Right Bank"
St-Émilion/Pomerol

GOOD VINTAGES	GREAT VINTAGES	OLDER GREAT VINTAGES
1995	1990	1982
1996	1998*	1989
1997	2000*	
1999	2001	
2002	2005*	
2003		
2004		
2006		

*Note: * signifies an exceptional year*

GRAND CRU CLASSÉ EN 1855

2000

CHATEAU LAGRANGE

SAINT-JULIEN

APPELLATION SAINT-JULIEN CONTRÔLÉE

CHÂTEAU LAGRANGE SA
PROPRIÉTAIRE À SAINT-JULIEN BEYCHEVELLE (GIRONDE) · FRANCE

Alc. 13% by vol.   MIS EN BOUTEILLE AU CHATEAU   e 750 ml

PRODUCE OF FRANCE · BORDEAUX

How do I buy—and drink—a red Bordeaux?

One of the biggest misconceptions about Bordeaux wines is that they are all very expensive. In reality, there are thousands of Bordeaux wines at all different price ranges.

First and foremost, ask yourself if you want to drink the wine now, or if you want to age it. A great château Bordeaux in a great vintage needs a *minimum* of ten years to age. Going down a level, a Cru Bourgeois or a second label of a great château in a great vintage needs a minimum of five years to age. A regional wine may be consumed within two or three years of the vintage year, while a wine labeled simply Appellation Bordeaux Contrôlée is ready to drink as soon as it's released.

The next step is to be sure the vintage is correct for what you want. If you're looking for a wine you want to age, you must look for a great vintage. If you want a wine that's ready to drink now and you want a greater château, you should choose a lesser vintage. If you want a wine that's ready to drink now and you want a great vintage, you should look for a lesser château.

In addition, remember that Bordeaux wines are a blend of grapes. Ask yourself if you're looking for a Merlot-style Bordeaux, such as St-Émilion or Pomerol, or if you're looking for a Cabernet style, such as Médoc or Graves, remembering that the Merlot is more accessible and easier to drink when young.

What separates a $20 red Bordeaux from a $300 red Bordeaux?

- The place the grapes are grown

- The age of the vines (usually the older the vine, the better the wine)

- The yield of the vine (lower yield means higher quality)

- The winemaking technique (for example, how long the wine is aged in wood)

- The vintage

ON DRINKING THE WINES OF BORDEAUX

"The French drink their Bordeaux wines too young, afraid that the Socialist government will take them away.

"The English drink their Bordeaux wines very old, because they like to take their friends down to their wine cellars with the cobwebs and dust to show off their old bottles.

"And the Americans drink their Bordeaux wines exactly when they are ready to be drunk, because they don't know any better."

—Author Unknown

HAD YOU dined at the Four Seasons restaurant in New York when it first opened in 1959, you could have had a 1918 Château Lafite-Rothschild for $18, or a 1934 Château Latour for $16. Or if those wines were a bit beyond your budget, you could have had a 1945 Château Cos d'Estournel for $9.50.

Is it necessary to pay a tremendous sum of money to get a great-tasting red Bordeaux wine?

It's nice if you have it to spend, but sometimes you don't. The best way to get the most for your money is to use what I call the reverse pyramid method. For example: Let's say you like Château Lafite-Rothschild, which is at the top of the pyramid at left ($$$$), but you can't afford it. What do you do? Look at the region. It's from Pauillac. You have a choice: You can go back to the 1855 classification and look for a fifth-growth wine from Pauillac that gives a flavor for the region at a lesser price ($$$), though not necessarily one-fifth the price of a first growth. Still too pricey? Drop a level further on the inverted pyramid and go for a Cru Bourgeois from Pauillac ($$). Your other option is to buy a regional wine labeled "Pauillac" ($).

I didn't memorize the seven thousand châteaux myself. When I go to my neighborhood retailer, I find a château I've never heard of. If it's from Pauillac, from a good vintage, and it's twenty to twenty-five dollars, I buy it. My chances are good. Everything in wine is hedging your bets.

Another way to avoid skyrocketing Bordeaux châteaux' prices is to take the time to look for their second-label wines. These wines are from the youngest parts of the vineyard and are lighter in style and quicker to mature but are usually a third the price of the château wine.

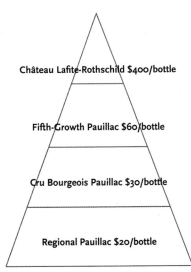

Château Lafite-Rothschild $400/bottle

Fifth-Growth Pauillac $60/bottle

Cru Bourgeois Pauillac $30/bottle

Regional Pauillac $20/bottle

CHÂTEAU	SECOND LABEL WINES
Château Lafite-Rothschild	Carruades de Lafite Rothschild
Château Latour	Les Forts de Latour
Château Haut-Brion	Bahans du Château Haut-Brion
Château Margaux	Pavillon Rouge du Château Margaux
Château Mouton-Rothschild	Petit Mouton
Château Léoville-Barton	La Réserve de Leoville Barton
Château Léoville-Las-Cases	Clos du Marquis
Château Pichon Lalande	Réserve de la Comtesse
Château Pichon Longueville	Les Tourelles de Pichon
Château Palmer	Réserve du Général
Château Lynch-Bages	Château Haut Bages-Averous

FOR FURTHER READING

I recommend *Grands Vins* by Clive Coates, M.W.; *The Bordeaux Atlas and Encyclopedia of Chateaux* by Hubrecht Duijker and Michael Broadbent; and *Bordeaux* by Robert M. Parker Jr.

SINCE 1882, WHEN the venerable French company Guerlain first produced a lip balm containing Bordeaux wine, nursing mothers have used it as a salve for chapped nipples. "It's a wonderfully soothing emollient, and red wine's tannic acid has healing properties," says Elisabeth Sirot, *attaché de presse* at Guerlain's Paris office. "Frenchwomen have always known this secret." Sirot used it when nursing all four of her children (she learned the tip from her own mother). How sensual—especially considering the American alternative is petroleum jelly.

WHILE NAPOLEON BONAPARTE preferred the Burgundy Chambertin, the late president Richard Nixon's favorite wine was Château Margaux. Nixon always had a bottle of his favorite vintage waiting at his table from the cellar of the famous "21" Club in New York.

WINE AND FOOD

DENISE LURTON-MOULLE *(Château La Louvière, Château Bonnet): With Château La Louvière Rouge: roast leg of lamb or grilled duck breast.*

JEAN-MICHEL CAZES *(Château Lynch-Bages, Château Haut-Bages-Averous, Château Les Ormes-de-Pez): "For Bordeaux red, simple and classic is best! Red meat, such as beef and particularly lamb, as we love it in Pauillac. If you can grill the meat on vine cuttings, you are in heaven."*

JACQUES AND FIONA THIENPONT *(Château Le Pin): Sunday lunches at Le Pin include lots of local oysters with chilled white Bordeaux followed by thick entrecôte steaks on the barbecue with shallots and a selection of the family's Pomerols, Margaux, or Côtes de France red Bordeaux.*

ANTONY PERRIN *(Château Carbonnieux): With red Bordeaux: magret de canard (duck breast) with wild mushrooms, or pintade aux raisins (Guinea hen with grapes).*

CHRISTIAN MOUEIX *(Château Pétrus): With red Bordeaux, especially Pomerol wine: Lamb is a must.*

MIS EN BOUTEILLE AU CHATEAU

CHATEAU
LES ORMES DE PEZ
SAINT-ESTÈPHE
1995

13.0%Vol · APPELLATION SAINT-ESTÈPHE CONTRÔLÉE · 750ml
A. CAZES, PROPRIETAIRE A SAINT ESTEPHE (FRANCE)
PRODUCE OF FRANCE

Questions for Class Five:
The Red Wines of Bordeaux

12. What percentage does the Grand Cru Classé Châteaux represent of the total volume of Bordeaux?   121

13. How many châteaux were classified in the official classification of the Mèdoc?   121

14. In what year were the wines of Graves first classified?   126

15. In what year were the wines of St. Émilion first classified?   127

16. What is the primary red grape used in the production of St. Émilion wine?   127

17. Name three great recent vintages from the "left bank" of Bordeaux.   128

18. Name three great recent vintages from the "right bank" of Bordeaux.   128

19. Name one château from each of the 5 growths: 123

   1st _____

   2nd _____

   3rd _____

   4th _____

   5th _____

20. Name two second-label wines of classified châteaux.   131

The Red Wines of California

❦

RED VS. WHITE • MAJOR RED GRAPES OF CALIFORNIA •

RED-GRAPE BOOM • MERITAGE • STYLES • TRENDS

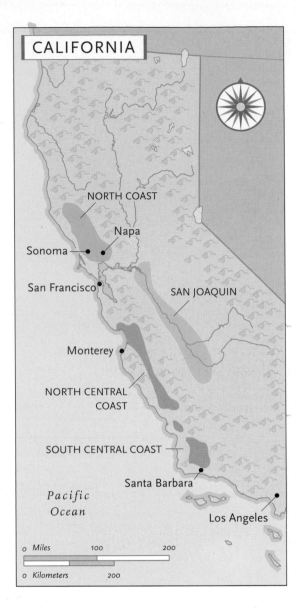

CALIFORNIA

NORTH COAST

Napa

Sonoma

San Francisco

SAN JOAQUIN

Monterey

NORTH CENTRAL
COAST

SOUTH CENTRAL COAST

Santa Barbara

*Pacific
Ocean*

Los Angeles

0 *Miles*   100   200

0 *Kilometers*   200

ACREAGE IN CALIFORNIA

Currently there are 290,233 acres in red grapes and 180,744 in white grapes planted in California.

More About California Wines

SINCE WE'VE ALREADY covered the history and geography of California in Class Two, it might be a good idea to go back and review the main viticultural areas of California wine country on page 51, before you continue with the red wines of California. Then, consider the following question that inevitably comes up in my class at the Windows on the World Wine School.

Are Americans drinking more white wine or red?

The chart below shows you the trend of wine consumption in the United States over the last thirty years. When I first began studying wines in 1970, people were more interested in red wine than white. From the mid-1970s, when I started teaching, into the mid-1990s, my students showed a definite preference for white wine. Fortunately for me (since I am a red wine drinker), the pendulum is surely swinging back to more red wine drinkers.

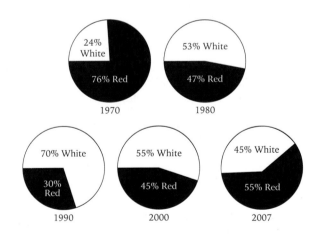

24% White

76% Red

1970

53% White

47% Red

1980

70% White

30% Red

1990

55% White

45% Red

2000

45% White

55% Red

2007

RED VS. WHITE—CONSUMPTION IN THE UNITED STATES

Why this change?

Looking back at the American obsession with health and fitness in the 1970s and 1980s, many people switched from meat and potatoes to fish and vegetables—a lighter diet that called more for white wine than red. "Chardonnay" became the new buzzword that replaced the call for "a glass of white wine." Bars that never used to stock wine—nothing decent, anyway—began to carry an assortment of fine wines by the glass, with Chardonnay, by far, the best-selling wine. Today, steak is back and the new buzzwords are Cabernet Sauvignon, Merlot, and Syrah.

Another major reason for the dramatic upturn in red wine consumption is the power of the media. Television popularized the so-called French Paradox (see box below).

Finally, perhaps the most important reason that red wine consumption has increased in the United States is that California is producing a much better quality red wine than ever before. One of the reasons for improved quality is the replanting of vines over the last twenty years due to the phylloxera problem. Some analysts thought the replanting would be financially devastating to the California wine industry, but in reality it may have been a blessing in disguise, especially with regard to quality.

The opportunity to replant allowed vineyard owners to increase their red grape production. It enabled California grape growers to utilize the knowledge they have gained over the years with regard to soil, climate, microclimate, trellising, and other viticultural practices.

Bottom line: California reds are already some of the greatest in the world, with more and better to come.

TOP THREE red grapes planted in California:
1. Cabernet Sauvignon
2. Merlot
3. Zinfandel

THERE ARE more than 30 wine grape varieties planted in California.

CALIFORNIA TOTALS IN 2007

Varietal	Acreage
Chardonnay	92,091
Cabernet	75,909
Merlot	51,570

ACCORDING TO the U.S. Dietary Guidelines:

5 oz. wine = 100 calories
12 oz. beer = 150 calories
1.5 oz. distilled spirits = 100 calories

WINE IS fat-free and contains no cholesterol.

FROM 1991 TO 2005, sales of red wine in the United States grew by more than 125%.

THE FRENCH PARADOX

In the early 1990s, the TV series 60 Minutes twice aired a report on a phenomenon known as the French Paradox—the fact that the French have a lower rate of heart disease than Americans, despite a diet that's higher in fat. Since the one thing the American diet lacks, in comparison to the French diet, is red wine, some researchers were looking for a link between the consumption of red wine and a decreased rate of heart disease. Not surprisingly, in the year following this report, Americans increased their purchases of red wines by 39 percent.

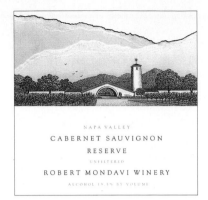

What are the major red grapes in California?

There are five major red grapes grown in California.

Cabernet Sauvignon: Considered the most successful red grape in California, it yields some of the greatest red wines in the world. Cabernet Sauvignon is the predominant variety used in the finest red Bordeaux wines, such as Château Lafite-Rothschild and Château Latour. Almost all California Cabernets are dry, and depending upon the producer and vintage, they range in style from light and ready to drink, to extremely full-bodied and long-lived. California Cabernet has become the benchmark for some of the best California wines.

My favorite California Cabernet Sauvignons are:

Arrowood	Joseph Phelps
Beaulieu Private Reserve	La Jota
Beringer Private Reserve	Laurel Glen
Cakebread	Mondavi Reserve
Caymus	Opus One
Chappellet	Paul Hobbs
Chateau Montelena	Pine Ridge
Chateau St. Jean, Cinq Cepages	Pride Mountain
Dalla Valle	Ridge Monte Bello
Diamond Creek	Schrader
Duckhorn	Shafer Hillside Select
Dunn Howell Mountain	Silver Oak
Gallo of Sonoma Estate	Spottswood
Groth Reserve	Staglin
Heitz	Stag's Leap Cask
Hess Collection	Whitehall Lane
Jordan	

BEST BETS FOR CABERNET SAUVIGNON (NAPA VALLEY)

1994*	1995*	1996*	1997*	1999*	2001*
2002*	2003	2004	2005*	2006*	2007

Note: * signifies exceptional vintage

Pinot Noir: Known as the "headache" grape because of its fragile quality, Pinot Noir is temperamental, high maintenance, expensive, and difficult to grow and make into wine. The great grape of the Burgundy region of France—responsible for such famous wines as Pommard, Nuits-St-Georges, and Gevrey-Chambertin—is also one of the principal grapes in French Champagne. In California, many years of experimentation in finding the right location to plant the Pinot Noir and to perfect the fermentation techniques have elevated some of the Pinot Noirs to the status of great wines. Pinot Noir is usually less tannic than Cabernet and matures more quickly, generally in two to five years. Because of the extra expense involved in growing this grape, the best examples of Pinot Noirs from California may cost more than other varietals. The three top counties for Pinot Noir are Sonoma (9,876 acres), Monterey (4,195 acres), and Santa Barbara (2,863 acres).

My favorite California Pinot Noirs are:

Acacia	Marcassin
Artesa	Merry Edwards
Au Bon Climat	Morgan
Calera	Patz & Hall
Cline	Paul Hobbs
Dehlinger	Robert Mondavi
Etude	Robert Sinskey
Flowers	Saintsbury
Gary Farrell	Sanford
J. Rochioli	Sea Smoke
Kosta Browne	Siduri
Littorai	Williams Selyem

BEST BETS FOR PINOT NOIR (NORTH COAST)

2001* 2002* 2003* 2004* 2005* 2006

*Note: * signifies exceptional vintage*

ONE AUTHOR, trying to sum up the difference between a Pinot Noir and a Cabernet Sauvignon, said, "Pinot is James Joyce, while Cabernet is Dickens. Both sell well, but one is easier to understand."

COMMON PINOT NOIR aromas:

Red berries	Red cherry
Leather	Tobacco (older)

SOUTHERN CALIFORNIA—especially the Santa Barbara area, as the characters in the movie *Sideways* could tell you—has become one of the prime locations for Pinot Noir production, with plantings up by more than 200% in the past decade. In fact, Pinot Noir sales overall have increased at least 20% since the film was released in 2004.

CARNEROS, MONTEREY, Sonoma, and the Anderson Valley are great places to grow Pinot Noir because of their cooler climate.

RED-GRAPE BOOM

Look at the chart below to see how many acres of the major red grapes were planted in California in 1970, and how those numbers have increased. Rapid expansion has been the characteristic of the California wine industry!

TOTAL BEARING ACREAGE OF RED-WINE GRAPES PLANTED
GRAPE-BY-GRAPE COMPARISON

GRAPE	1970	1980	1990	2007
Cabernet Sauvignon	3,200	21,800	24,100	75,909
Merlot	100	2,600	4,000	51,570
Zinfandel	19,200	27,700	28,000	49,697
Pinot Noir	2,100	9,200	8,600	24,188
Syrah			400	17,918

SOME EXAMPLES of Meritage wines of California:

Dominus (Christian Moueix)
Insignia (Phelps Vineyards)
Magnificat (Franciscan)
Opus One (Mondavi/Rothschild)
Cain Five
Trefethen Halo

What are Meritage wines?

Meritage (which rhymes with *heritage*) is the name for red and white wines made in America from a blend of the classic Bordeaux wine-grape varieties. This category was created because many winemakers felt stifled by the required minimum amount (75 percent) of a grape that must go into a bottle for it to be named for that variety. Some winemakers knew they could make a better wine with a blend of, say, 60 percent of the major grape and 40 percent of secondary grapes. This blending of grapes allows producers of Meritage wines the same freedom that Bordeaux winemakers have in making their wines.

For red wine, the varieties include Cabernet Sauvignon, Merlot, Cabernet Franc, Petit Verdot, and Malbec. For white wine, the varieties include Sauvignon Blanc and Sémillon.

When I buy a Cabernet, Zinfandel, Merlot, Pinot Noir, Syrah, or Meritage wine, how do I know which style I'm getting? Is the style of the wine indicated on the label?

Unless you just happen to be familiar with a particular vineyard's wine, you're stuck with trial-and-error tastings. You're one step ahead, though, just by knowing that you'll find drastically different styles from the same grape variety.

With some 1,900 wineries in California and more than half of them

OPUS ONE

Amidst grand hoopla in the wine world, Robert Mondavi and the late Baron Philippe de Rothschild released Opus One. "It isn't Mouton and it isn't Mondavi," said Robert Mondavi. Opus One is a Bordeaux-style blend made from Cabernet Sauvignon, Merlot, and Cabernet Franc grapes grown in Napa Valley. It was originally produced at the Robert Mondavi Winery in Napa Valley, but is now produced across Highway 29 in its own spectacular winery.

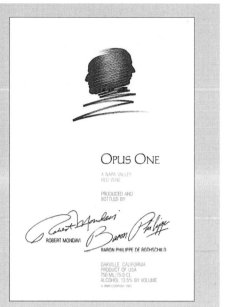

TROPHY HUNTING

Wine collectors started a frenzy by buying Cabernet Sauvignon from small California wineries at extraordinary prices. These "cult" wineries produce very little wine—with hefty price tags.

Araujo	4,000 cases
Colgin Cellars	2,500 cases
Harlan Estate	1,800 cases
Bryant Family	1,000 cases
Dalla Valle	1,000 cases
Screaming Eagle	500 cases
Grace Family	350 cases

producing red wines, it is virtually impossible to keep up with the ever-changing styles that are being produced. One of the recent improvements in labeling is that more wineries are adding important information to the back label indicating when the wine is ready to drink, if it should be aged, and many even offer food suggestions.

To avoid any unpleasant surprises, I can't emphasize enough the importance of an educated wine retailer. One of the strongest recommendations I give—especially to a new wine drinker—is to find the right retailer, one who understands wine and your taste.

Do California red wines age well?

Absolutely, especially from the best wineries that produce Cabernet Sauvignon and Zinfandel. I have been fortunate to taste some early examples of Cabernet Sauvignon going back to the 1930s, 1940s, and 1950s, which for the most part were drinking well—some of them were outstanding—proving to me the longevity of certain Cabernets. Zinfandels and Cabernet Sauvignons from the best wineries in great vintages will need a minimum of five years before you drink them, and they will get better over the next ten years. That's at least fifteen years of great enjoyment.

ONE OF THE most memorable tastings I have ever attended in my career was for the fiftieth anniversary of Beaulieu's Private Reserve wine. Over a two-day period, we tasted every vintage from 1936 to 1986 with winemaker André Tchelistcheff. I think everyone who attended the tasting was amazed and awed by how well many of these vintages aged.

NAPA VALLEY—40 YEARS LATER: 1968–2008

1968: $2,000–$4,000 (cost per acre)
2008: $150,000–$350,000 (cost per acre)

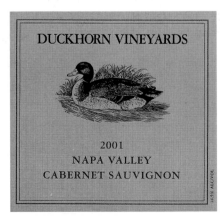

However, one of the things I have noticed in the last ten years, not only tasting as many California wines as I have, but also tasting so many European wines, is that California wines seem to be more accessible when young, as opposed to, say, a Bordeaux. I believe this is one of the reasons California wines sell so well in the United States, especially in restaurants.

What have been the trends in the red wines of California over the last fifteen years?

To best answer that question, we should go back even further to see where the trends have been going for the last thirty years or so. The 1960s were a decade of expansion and development. The 1970s were a decade of growth, especially in terms of the number of wineries that were established in California and the corporations and individuals that became involved. The 1980s and 1990s were the decades of experimentation, in grape growing as well as in wine-making and marketing techniques.

Over the past ten years, I have seen the winemakers finally get a chance to step back and fine-tune their wine. Today, they are producing wines that have tremendous structure, finesse, and elegance that many lacked in the early years of the California winemaking renaissance. They are also making wines that can give pleasure when young, and also great wines that I hope I will be around to share with my grandchildren. The benchmark for quality has increased to such a level that the best wineries have gotten better, but more important to the consumer, even the wines under twenty dollars are better than ever before.

Though California winemakers have settled down, they have not given up experimentation altogether, if you consider the many new grape varieties coming out of California these days. I expect to see more wines made with grapes such as the Mourvèdre, Grenache, Sangiovese, and especially Syrah, continuing the trend toward diversity in California red wines.

Red wines are no longer the sole domain of Napa and Sonoma. Many world-class reds are being produced in the Central Coast regions of California such as Monterey and Santa Barbara, San Luis Obispo, and Santa Clara.

FOR FURTHER READING

I recommend *The Wine Atlas of California* by James Halliday; *Making Sense of California Wine* and *New California Wine* by Matt Kramer; *Wine Spectator's California Wine* by James Laube; and *The Wine Atlas of California and the Pacific Northwest* by Bob Thompson.

WINE AND FOOD

MARGRIT BIEVER AND ROBERT MONDAVI: *With Cabernet Sauvignon: lamb, or wild game such as grouse and caribou. With Pinot Noir: pork loin, milder game such as domestic pheasant, coq au vin.*

TOM JORDAN: *"Roast lamb is wonderful with the flavor and complexity of Cabernet Sauvignon. The wine also pairs nicely with sliced breast of duck, and grilled squab with wild mushrooms. For a cheese course with mature Cabernet, milder cheeses such as young goat cheeses, St. André and Taleggio, are best so the subtle flavors of the wine can be enjoyed."*

MARGARET AND DAN DUCKHORN: *"With a young Merlot, we recommend lamb shanks with crispy polenta, or grilled duck with wild rice in Port sauce. One of our favorites is barbecued leg of lamb with a mild, spicy fruit-based sauce. With older Merlots at the end of the meal, we like to serve Cambazzola cheese and warm walnuts."*

JANET TREFETHEN: *With Cabernet Sauvignon: prime cut of well-aged grilled beef; also—believe it or not— with chocolate and chocolate-chip cookies. With Pinot Noir: roasted quail stuffed with peeled kiwi fruit in a Madeira sauce. Also with pork tenderloin in a fruity sauce.*

PAUL DRAPER *(Ridge Vineyards): With Zinfandel: a well-made risotto of Petaluma duck. With aged Cabernet Sauvignon: Moroccan lamb with figs.*

WARREN WINIARSKI *(Stag's Leap Wine Cellars): With Cabernet Sauvignon: lamb or veal with a light sauce.*

JOSH JENSEN *(Calera Wine Co.): "Pinot Noir is so versatile, but I like it best with fowl of all sorts—chicken, turkey, duck, pheasant, and quail, preferably roasted or mesquite grilled. It's also great with fish such as salmon, tuna, and snapper."*

RICHARD ARROWOOD: *With Cabernet Sauvignon: Sonoma County spring lamb or lamb chops prepared in a rosemary herb sauce.*

DAVID STARE *(Dry Creek Vineyard): "My favorite food combination with Zinfandel is marinated, butterflied leg of lamb. Have the butcher butterfly the leg, then place it in a plastic bag. Pour in half a bottle of Dry Creek Zinfandel, a cup of olive oil, six mashed garlic cloves, salt and pepper to taste. Marinate for several hours or overnight in the refrigerator. Barbecue until medium rare. While the lamb is cooking, take the marinade, reduce it, and whisk in several pats of butter for thickness. Yummy!"*

BO BARRETT *(Chateau Montelena Winery): With Cabernet Sauvignon: a good rib eye, barbecued with a teriyaki-soy-ginger-sesame marinade; venison or even roast beef prepared with olive oil and tapenade with rosemary, or even lamb. But when it comes to a good Cabernet Sauvignon, Bo is happy to enjoy a glass with "nothing at all—just a good book."*

PATRICK CAMPBELL *(owner/winemaker, Laurel Glen Vineyard): "With Cabernet Sauvignon, try a rich risotto topped with wild mushrooms."*

JACK CAKEBREAD: *"I enjoy my 1994 Cakebread Cellars Napa Valley Cabernet Sauvignon with farm-raised salmon with a crispy potato crust or an herb-crusted Napa Valley rack of lamb, with mashed potatoes and a red-wine sauce."*

ED SBRAGIA *(winemaker, Beringer Vineyards): "I like my Cabernet Sauvignon with rack of lamb, beef, or rare duck."*

TOM MACKEY *(winemaker, St. Francis Merlot): With St. Francis Merlot Sonoma County: Dungeness crab cakes, rack of lamb, pork roast, or tortellini. With St. Francis Merlot Reserve: hearty minestrone or lentil soup, venison, or filet mignon, or even a Caesar salad.*

Questions for Class Six:
The Red Wines of California

8. Which county in California has the most plantings of Pinot Noir?

9. In which two French wine regions would you find Pinot Noir?

10. What grape in California has a similar DNA as the Italian grape Primitivo?

11. In which French wine region would you find Syrah?

12. In what counties are most of the Syrah grapes planted?

13. What was the most planted red grape in California in 1970?

14. What is a Meritage wine?

15. Name two Meritage wines.

Wines of the World: Italy, Spain, Australia, Chile, and Argentina

ITALY · DOC · TUSCANY · PIEDMONT · VENETO · TRENDS · SPAIN · RIOJA ·

AUSTRALIA · HOW TO READ AN AUSTRALIAN LABEL · CHILE · ARGENTINA

The Red Wines of Italy

ITALY PRODUCES 50% white wines and 50% red.

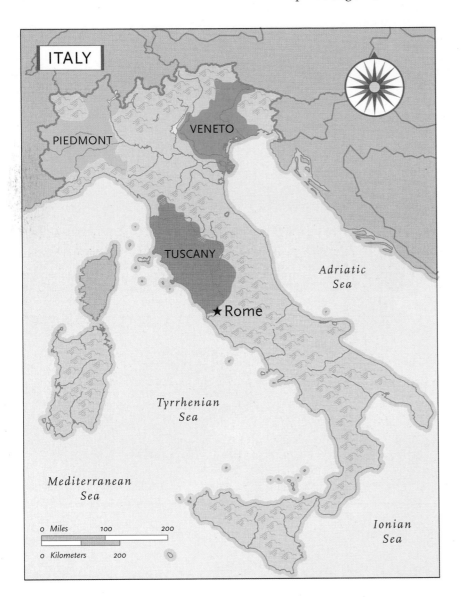

ITALY IS THE WORLD'S second largest producer of wine after France. It has been producing wine for more than three thousand years, and the vines grow everywhere. As one retailer of fine Italian wine once told me, "There is no country. Italy is one vast vineyard from north to south."

Italian wines are good for any occasion—from quaffing to serious tasting. Some of my favorite wines are Italian. In fact, 25 percent of my personal wine cellar is stocked with them.

There are more than two thousand different wine labels, if you care to memorize them, twenty regions, and ninety-six provinces. But don't worry. If you want to know the basics of Italian wines, concentrate on the three regions listed below, and you'll be well on your way to having Italy in the palm of your hand.

TUSCANY

PIEDMONT

VENETO

What are the major red-grape varieties in Italy?

There are hundreds of indigenous grapes planted throughout Italy. In Tuscany, the major red-grape variety is Sangiovese, and in Piedmont it is Nebbiolo.

How are Italian wines controlled?

The Denominazione di Origine Controllata (abbreviated DOC), the Italian equivalent of the French AOC, controls the production and labeling of the wine. Italy's DOC laws went into effect in 1963.

In 1980 the Italian wine board took quality control even one step beyond the regular DOC, when it added the higher-ranking DOCG. The G stands for *Garantita,* meaning that, through tasting-control boards, they absolutely guarantee the stylistic authenticity of a wine.

As of 2008, the wines from Piedmont and Tuscany that qualified for the DOCG were:

TUSCANY	PIEDMONT
Vernaccia di San Gimignano	Moscato d'Asti/Asti
Chianti	Gattinara
Chianti Classico	Barbaresco
Vino Nobile di Montepulciano	Barolo
Carmignano Rosso	Acqui or Brachetto d'Acqui
Brunello di Montalcino	Ghemme
Morellino di Scansano	Gavi or Cortese di Gavi
	Roero
	Dolcetto di Dogliani Superiore

Nine of the thirty-five Italian DOCG wines are from Piedmont and seven are from Tuscany. More than one-third of the DOCGs are located in these two regions, which is why your study of Italian wines should focus on them.

DOC LAWS

The DOC governs:

1. *The geographical limits of each region*

2. *The grape varieties that can be used*

3. *The percentage of each grape used*

4. *The maximum amount of wine that can be produced per acre*

5. *The minimum alcohol content of the wine*

6. *The aging requirements, such as how long a wine should spend in wood or bottle, for certain wines*

MANY WINE producers in Italy are now making wines from Cabernet Sauvignon, Merlot, and Chardonnay.

TOP THREE REGIONS IN PRODUCTION OF ITALIAN WINES

1.	Veneto	17.7%
2.	Piedmont	17.1%
3.	Tuscany	10.7%

MY ITALIAN-WINE friends sometimes refer to Veneto as Tri-Veneto, which includes Trentino, Alto-Aldige, and Friuli. Some of the best white wines of Italy come from those regions.

AT PRESENT, there are 35 wines that are entitled to the DOCG designation.

DOCG WINES from regions other than Tuscany and Piedmont include Taurasi, Greco di Tufo, and Fiano di Avellino from Campania; Albana di Romagna from Emilia-Romagna; Torgiano Riserva and Montefalco Sagrantino di Montefalco from Umbria; Franciacorta, Valtellina Superiore, and Sforzato di Valtellina from Lombardy; Recioto di Soave, Bardolino Superiore, and Soave Superiore from Veneto; Vermentino de Gallura from Sardinia; Ramandolo and Colli Orientale del Fruili Picolit from Friuli-Venezia Giulia; Montelpulciano d'Abruzzo Colline Teramane from Abruzzi; Conero and Vernaccia di Serrapetrona from Marches; and Cerasvolo di Vittoria from Sicily.

THERE ARE more than 300 DOC wines accounting for 20% of Italy's total wine production.

OF ALL Italian DOC wines, 60% are red.

THE BIGGEST difference between the AOC of France and the DOC of Italy is that the DOC has aging requirements.

ONLY ONE-FIFTH of all Chianti is Chianti Classico Riserva.

SUPER TUSCAN

As in California, some winemakers in Italy wanted to experiment with grape varieties and blends beyond what was permitted by the DOC regulations, so they decided to produce their own styles of wine. In Tuscany, these wines have become known as the "Super Tuscans." Among the better known of these proprietary Italian wines are Sassicaia, Tignanello, Ornellaia, Cabreo Il Borgo, Solaia, Summus, and Excelsus.

TUSCANY—THE HOME OF CHIANTI

Why did Chianti have such a bad image until recently?

One reason was the little straw-covered flasks (*fiaschi*) that the wine was bottled in—nice until restaurants hung the bottles from the ceiling next to the bar, along with the sausage and the provolone, or made them into candle holders for the table. So Chianti developed an image as a cheap little red wine to be bought for five dollars a jug.

My own feeling is that Chianti Classico Riserva is one of the best values in Italian wine today.

What are the different levels of Chianti?

Chianti ($)—The first level.
Chianti Classico ($$)—From the inner historic district of Chianti.
Chianti Classico Riserva ($$$$)—From a Classico area, and must be aged for a minimum of two years, three months.

How should I buy Chianti?

First of all, find the style of Chianti you like best. There is a considerable variation in Chianti styles. Second, always buy from a shipper or producer whom you know—one with a good, reliable reputation. Some quality Chianti producers are:

ANTINORI	FRESCOBALDI
BADIA A COLTIBUONO	MELINI
BROLIO	MONSANTO
CASTELLO BANFI	NOZZOLE
CASTELLO DI AMA	RICASOLI
FONTODI	RUFFINO

Which grapes are used in Chianti?

According to updated DOCG requirements, winemakers are required to use at least 80 percent Sangiovese to produce Chianti. The DOCG also encourages the use of other grapes by allowing an unprecedented 20 percent nontraditional grapes (Cabernet Sauvignon, Merlot, Syrah, etc.). These changes, along with better winemaking techniques and better vineyard development, have all contributed greatly to improving Chianti's image over the last twenty years. A separate DOCG has been established for Chianti Classico, and many producers of this wine now use 100 percent Sangiovese.

Which other high-quality wines come from Tuscany?

Three of the greatest Italian red wines are Brunello di Montalcino, Vino Nobile di Montepulciano, and Carmignano. If you purchase the Brunello, keep in mind that it probably needs more aging (five to ten years) before it reaches peak drinkability. There are more than 150 producers of Brunello. My favorite producers of Brunello are: Carpineto, Marchesi de Frescobaldi, Barbi, Altesino, Castelgiorondo, La Poderina, Il Poggione, Lisini, La Fuga, Poggio Antico, Col d'Orcia, Castello Banfi, Caparzo, Gaja, and Soldera. Those of Vino Nobile are: Avignonesi, Boscarelli, Fassati, Dei, Fattoria del Cerro, and Poggio alla Sala. For Carmignano, look for Villa Capezzana, Poggiolo, and Artimino.

BEST BETS FOR WINES FROM TUSCANY

1997**	1999**	2001*	2003*
2004*	2005*	2006**	2007*

*Note: * signifies exceptional vintage   ** signifies extraordinary vintage*

PIEDMONT—THE BIG REDS

Some of the finest red wines are produced in Piedmont. Two of the best DOCG wines to come from this region in northwest Italy are Barolo and Barbaresco.

The major grapes of Piedmont are:

DOLCETTO
BARBERA
NEBBIOLO

Barolo and Barbaresco, the "heavyweight" wines from Piedmont, are made from the Nebbiolo variety. These wines have the fullest style and a high alcohol content. Be careful when you try to match young vintages of these wines with your dinner; they may overpower the food.

BRUNELLO DI MONTALCINO, because of its limited supply, is sometimes very expensive. For one of the best values in Tuscan red wine, look for Rosso di Montalcino.

BRUNELLO IS CHANGING

Beginning with the 1995 vintage, Brunellos are required to be aged in oak for a minimum of two years instead of the previous three. Result? A fruitier, more accessible wine.

THREE REASONS to visit Piedmont in the fall: the harvest, the food, and the white truffles.

APPROXIMATELY 50% of the vineyards in Piedmont are Barbera.

MORE THAN 90% of the Piedmont grapes are planted on hillsides.

BOTH THE 2002 AND 2003 vintages in Italy were well below normal production levels, and the 2002 vintage is the smallest harvest in 10 years.

OLDER GREAT vintages of Piedmont: 1982, 1985, 1988, 1989

THE PIEDMONT region had six great vintages in a row: 1996–2001.

THE 2002 VINTAGE in Piedmont lost most of its grapes to a September hailstorm.

PIEDMONT'S PRODUCTION

65% red
18% spumante (sparkling white)
17% white

THE PRODUCTION of Barolo and Barbaresco at 11 million bottles a year is equivalent to that of only a medium-size California winery!

IT IS SAID that when you begin drinking the red wines of Piedmont, you start with the lighter-style Barbera and Dolcetto, move on to the fuller-bodied Barbaresco, until finally you can fully appreciate a Barolo. As the late vintner Renato Ratti said, "Barolo is the wine of arrival."

IN ONE of the biggest changes in the DOCG regulations, the wines of Barolo now have to be aged in wood for only one year, and the minimum alcohol has been changed to 12.5%. Before 1999, Barolo had a mandatory two years of wood aging and 13% minimum alcohol.

ANOTHER GREAT Piedmont wine is called Gattinara. Look for the Antoniolo Reservas.

My favorite producers of Piedmont wines are: Antonio Vallana, Fontanafredda, Gaja, Pio Cesare, Prunotto, Renato Ratti, Ceretto, G. Conterno, Robert Voerzio, Domenico Clerico, A. Conterno, M. Chiarlo, B. Giacosa, Marchesi di Gresy, Marchesi di Barolo, Vietti, Marcarini, Sandrone, Conterno Fantina, and Produttori d'Barbaresco.

BEST BETS FOR WINES FROM PIEDMONT

1990*	1996**	1997*	1998*	1999*	2000**
2001**	2003	2004**	2005*	2006*	2007

Note: * signifies exceptional vintage   ** signifies extraordinary vintage

BAROLO VS. BARBARESCO

BAROLO (MORE THAN 8 MILLION BOTTLES)	BARBARESCO (NEARLY 3 MILLION BOTTLES)
Nebbiolo grape	Nebbiolo grape
Minimum 12.5% alcohol	Minimum 12.5% alcohol
More complex flavor, more body	Lighter; sometimes less body than Barolo, but fine and elegant
Must be aged at least three years (one in wood)	Requires two years of aging (one in wood)
"Riserva" = five years of aging	"Riserva" = four years of aging

VENETO—THE HOME OF AMARONE

This is one of Italy's largest wine-producing regions. Even if you don't recognize the name immediately, I'm sure you've had Veronese wines at one time or another, like Valpolicella, Bardolino, and Soave. All three are very consistent, easy to drink, and ready to be consumed whenever you buy them. They don't fit into the category of a Brunello di Montalcino or a Barolo, but they're very good table wines and they're within everyone's budget. The best and most improved of the three is Valpolicella. Look for Valpolicella Superiore made by the *ripasso* method.

Easy-to-find Veneto producers are Bolla, Folonari, and Santa Sofia. Harder to find and higher-priced but worth it come from Allegrini, Anselmi, and Quintarelli.

THE NAME "Amarone" derives from *amar*, meaning "bitter," and *one* (pronounced "oh-nay"), meaning "big."

THE TOP five wines imported to the United States from Italy are:
1. Riunite
2. Casarsa
3. Bolla
4. Cavit
5. Ecco Domani

The above wines equal 43% of all imported table wine in the United States.

Ripasso is the adding back of grape skins from Amarone wine to Valpolicella, giving it extra alcohol and more flavor.

What's Amarone?

Amarone is a type of Valpolicella wine made by a special process in the Veneto region. Only the ripest grapes (Corvina, Rondinella, and Molinara) from the top of each bunch are used. After picking, they're left to "raisinate" (dry and shrivel) on straw mats. Does this sound familiar to you? It should, because this is similar to the process used to make German Trockenbeerenauslese and French Sauternes. One difference is that with Amarone, the winemaker ferments most of the sugar, bringing the alcohol content to 14 to 16 percent.

My favorite producers of Amarone are Masi, Bertani, Allegrini, Quintarelli, and Tommasi.

CLASSICO: All the vineyards are in the historical part of the region.

SUPERIORE: Higher levels of alcohol and longer aging.

BEST BETS FOR AMARONE

1990* 1993 1995* 1996 1997* 1998
2000* 2001 2002* 2003* 2005*

*Note: * signifies exceptional vintage*

HOW ITALIAN WINES ARE NAMED

Winemaking regions have different ways of naming their wines. In California, you look for the grape variety on the label. In Bordeaux, most often you will see the name of a château. But in Italy, there are three different ways that wine is named: by grape variety, village or district, or simply by a proprietary name. See the examples below.

GRAPE VARIETY	VILLAGE OR DISTRICT	PROPRIETARY
Barbera	Chianti	Tignanello
Nebbiolo	Barolo	Sassicaia
Sangiovese	Montalcino	Summus

What have been the trends in the red wines of Italy over the last two decades?

Going back a little further, we can see that as recently as twenty-five years ago most Italian wine was made to be consumed in Italy, and not for the export market. As one wine producer commented, "They didn't drink the wine, they ate the wine." To the Italians, wine was an everyday thing like salt and pepper on their table to enhance the taste of their food.

But over the last thirty years, winemaking has become more of a business, and the Italian winemakers' philosophy has changed considerably from making casual-drinking wines to much better-made wines that are also much more marketable around the world. They've accomplished this by using modern technology, modern vinification procedures, and updated vineyard management as a basis for experimentation.

Another area of major experimentation is with nontraditional grape varieties such as Cabernet Sauvignon and Merlot. One of the newest trends is single-vineyard labeling. As a result, the biggest news in the whole wine industry over the last thirty years is the change in Italian wines. When I talk about experimentation, you must remember that this isn't California we're talking about, but Italy, with thousands of years of traditions that are being changed. In Italy, the producers have had to unlearn and relearn winemaking techniques in order to make better wines.

The prices of Italian wines have also increased tremendously over the last ten years, which may be good news for the Italian wine producers (in that it enhances the image of their wines), but it isn't such good news for consumers.

Some of the wines from Italy have become among the most expensive in the world. That's not to say they're not worth it, but the pricing situation isn't the same as it was twenty years ago.

Beyond the best-known regions of Tuscany, Piedmont, and Veneto, many of the other twenty regions in Italy—especially Friuli and Trentino in the north and places like Umbria, Campania, Basilicata, Apulia, and Sicily in the south—are producing better and better wines, many of them made with indigenous grapes (Vermentino, Fiano, Negramora, etc.).

IN THE last 10 years, Italians have become more weight- and health-conscious, so they're changing their eating habits. As a result, the leisurely four-hour lunch and siesta is a thing of the past. Yes, all good things must come to an end.

"When you're having Italian wines, you must not taste the wine alone. You must have them with food."
—GIUSEPPE COLLA *of Prunotto*

Italian Whites

I AM OFTEN ASKED WHY I don't teach a class on Italian white wines. The answer is quite simple. Take a look at the most popular white wines: Soave, Frascati, and Pinot Grigio, among others. Most of them retail for less than fifteen dollars. The Italians traditionally do not put the same effort into making their white wines as they do their reds—in terms of style or complexity—and they are the first to admit it.

Plantings of international white varieties such as Chardonnay and Sauvignon Blanc, along with some of the better indigenous grapes, have recently elevated the quality of Italian white wines.

FOR GREAT Italian whites, try Gavi from Piedmont, and wines from the Friuli region.

PINOT GRIGIO is a white-grape variety that is also found in Alsace, France, where it is called Pinot Gris. It is also grown with success in Oregon and California.

IS PINOT GRIGIO HOT?

In 1970, Italy produced only 11,000 cases of Pinot Grigio, and in 2007, Pinot Grigio was the number one imported Italian varietal in the United States, with more than 10 million cases sold.

FOR FURTHER READING

I recommend *The Pocket Guide to Italian Wines* and *Wine Atlas of Italy* by Burton Anderson; *Vino Italiano* by Joseph Bastianich and David Lynch; *Italian Wine* by Victor Hazan; and *Italian Wines for Dummies* by Mary Ewing Mulligan and Ed McCarthy.

WINE AND FOOD

In Italy, the wine is made to go with the food. No meal is served without wine. Take it from the experts.

The following food-and-wine suggestions are based on what some of the Italian wine producers enjoy having with their wine. You don't have to take their word for it. Get yourself a bottle of wine, a tasty dish, and mangia!

AMBROGIO FOLONARI *(Ruffino):* *"Chianti with prosciutto, chicken, pasta, and of course pizza."* When it comes to a *Chianti Classico Riserva*, *Dr. Folonari says, "Pair it with a hearty prime-rib dinner or a steak."*

EZIO RIVELLA *(Castello Banfi):* *"A Chianti is good with all meat dishes, but I save the Brunello for 'stronger' dishes, such*

Continued . . .

"Piedmontese wines show better with food than in a tasting."

—ANGELO GAJA

GIUSEPPE COLLA of Prunotto offers his "Best Bets" in the form of advice. His general rule: In a good vintage, set a Barbaresco aside for a minimum of four years before drinking. In the same situation, put away a Barolo for six years. However, in a great vintage year, lay down a Barbaresco for six years and a Barolo for eight years. As they say, "Patience is a virtue"—especially with wine.

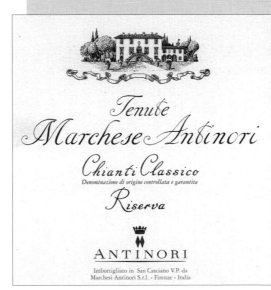

as steak, wild boar, pheasant, and other game, as well as Pecorino Toscano cheese."

ANGELO GAJA: *"Barbaresco with meat and veal, and also with mature cheeses that are 'not too strong,' such as Emmenthaler and Fontina."* Mr. Gaja advises against Parmesan and goat cheese when you have a Barbaresco. And if you're having a Barolo, Mr. Gaja's favorite is roast lamb.

GIUSEPPE COLLA *(Prunotto): "I enjoy light-style Dolcetto with all first courses and all white meat—chicken and veal especially."* He prefers not to have Dolcetto with fish. *"The wine doesn't stand up well to spicy sauce, but it's great with tomato sauce and pasta."*

RENATO RATTI: *Mr. Ratti once told me that both Barbera and Dolcetto are good with chicken and lighter foods. However, Barolo and Barbaresco need to be served with heavier dishes to match their own body. Mr. Ratti also suggests: a roast in* its natural sauce or, better yet, brasato al Barolo—braised in Barolo; meat cooked with wine; pheasant, duck, wild rabbit; and cheeses. For a special dish, try risotto al Barolo (rice cooked with Barolo wine). And when serving wine with dessert, Mr. Ratti recommends "strawberries or peaches with Dolcetto wine." The dryness in the wine, contrasted with the natural sweetness of the fruit, makes for a taste sensation!

LORENZA DE'MEDICI *(Badia a Coltibuono): Since Tuscan cooking is very simple, she recommends "an assortment of simple foods."* She prefers herbs to heavy sauces. With young Chianti, she suggests "roast chicken, squab, or pasta with meat sauce." To complement an older Chianti, she recommends a wide pasta with meat braised in Chianti, pheasant or other game, wild boar, or roast beef.

PIERO ANTINORI: *"I enjoy Chianti with the grilled foods for which Tuscany is famous, especially its bistecca alla Fiorentina (steak)."* He suggests poultry and even hamburgers as other tasty possibilities. With Chianti Classico Riserva, Mr. Antinori enjoys having the best of the vintages with wild boar and fine aged Parmesan cheese. *"The wine is a perfect match for roast beef, roast turkey, lamb, or veal."*

SPAIN

Atlantic Ocean

RIBERA DEL DUERO   RIOJA

PENEDÉS

PRIORAT   Barcelona

★ Madrid

Sevilla

Mediterranean Sea

SHERRY

SALES OF Spanish wine in the United States grew by 16% in 2005.

ALSO, look for the great rosés of the Navarra region.

SPAIN has the most vineyard acreage of any country in the world, with over 600 different varieties of grapes planted.

OTHER WINE-PRODUCING AREAS OF SPAIN

Rias Baixas
Toro
Jumilla

The Wines of Spain

THE MAIN WINEMAKING regions in Spain are:

LA RIOJA   CATALUÑA (PENEDÉS AND PRIORAT)

RIBERA DEL DUERO   THE SHERRY DISTRICT

We'll put aside the region of Sherry for now, since you'll become a Sherry expert in the next chapter. Let's start with Rioja, which is located in northern Spain, very near the French border.

In fact, it's less than a five-hour (two hundred–mile) drive to Bordeaux from Rioja, so it's no coincidence that Rioja wines often have a Bordeaux style. Back in the 1800s, many Bordeaux wine producers brought their expertise to this region.

MARQUÉS DE CÁCERES is owned by a Spaniard, who also owns Château Camensac, a fifth-growth Bordeaux.

VIÑOS DE PAGOS means that the wine comes from a single estate.

GRENACHE IS the French spelling of Garnacha, a Spanish grape, which was brought to France from Spain during the time of the popes of Avignon, some of whom were Spanish.

THE VIURA is the most used grape in white Rioja.

IN RIOJA, 80% of the grapes are red.

ALTHOUGH THERE are nearly 800 different wineries in Spain, about 80% of their production comes from a handful of companies.

THE NEW RIOJA

A new style of wine is emerging in Rioja. It is a contemporary wine that is bigger and more concentrated. Wineries of this style include Allende, Remelluri, Palacios, Remondo, and Remirez de Ganuza.

Why would a Frenchman leave his château in Bordeaux to go to a Spanish bodega in Rioja?

I'm glad you asked. It so happens that Frenchmen did travel from Bordeaux to Rioja at one point in history. Do you remember phylloxera? (See Prelude to Wine.) It's a plant louse that at one time killed nearly all the European vines and almost wiped out the Bordeaux wine industry.

Phylloxera started in the north and moved south. It first destroyed all the vines in Bordeaux, and some of the Bordeaux vineyard owners decided to establish vineyards and wineries in the Rioja district. It was a logical place for them to go because of the similar climate and growing conditions. The influence of the Bordelaise is sometimes apparent even in today's Rioja wines.

Which grapes are used in Rioja wines?

The major red grapes used in Rioja wine are:

TEMPRANILLO

GARNACHA

These grapes are blended with others to give Rioja wines their distinctive taste. On some labels you will see the word *cosecha*, which in Spanish means "harvest" or "vintage."

SPAIN'S WINE RENAISSANCE

Spain is the world's third largest producer of wine, behind Italy and France. Spain has more land dedicated to vines than any other country: 4.5 million acres!

Spain is also a country with a rich winemaking tradition. Many types of wine are produced: Cava (Méthode Champenoise sparkling wines) are produced in the Penedés region not far from Barcelona. Red wines and rosés are produced throughout the entire country. Fortified wines are produced mainly in the south. The best-known region for such wines is Jerez (Sherry), as you will see in Class Eight.

A twentieth-century renaissance extended nationwide to Spain's wine industry, where tremendous investments have been made throughout the country in viticulture and winemaking equipment. In Rioja alone, the number of wineries has increased from 42 to more than 500 since 1982.

Why are Rioja wines so easy to understand?

All you need to know when buying a Rioja wine is the style (level) and the reputation of the Rioja winemaker/shipper. The grape varieties are not found on the wine labels, and there's no classification to be memorized. The three major levels of Rioja wines are:

Crianza ($)—Released after two years of aging, with a minimum of one year in oak barrels.

Reserva ($$)—Released after three years of aging, with a minimum of one year in oak barrels.

Gran Reserva ($$$$)—Released after five to seven years of aging, with a minimum of two years in oak barrels.

THREE NEW CATEGORIES OF AGE FOR SPANISH WINES

Noble	12 months in barrel
Anejo	24 months in barrel
Viejo	36 months in barrel

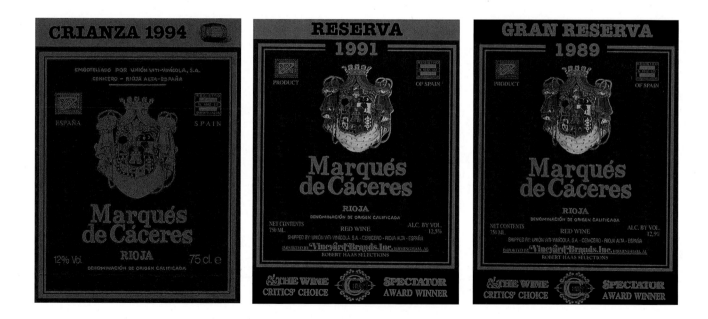

How would I know which Rioja wine to buy in the store?

You mean besides going with your preferred style and the reputation of the winemaker/shipper? You may also be familiar with a Rioja wine by its proprietary

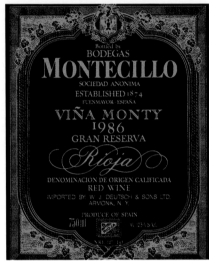

MARQUÉS DE MURRIETA was the first commercial bodega in Rioja, established in 1850.

SOME OF the top Rioja winemakers are saying that the 2001 and 2004 vintages are the best they have ever tasted.

RIBERA DEL DUERO wines are made from the Tinto Fino grape, a close cousin of Rioja's Tempranillo. The laws also allow the use of Cabernet Sauvignon, Malbec, Merlot, and small amounts of a white grape called Albillo.

IN THE LAST 25 years, Ribera del Duero has gone from 6 Bodegas to over 200.

name. The following are some bodegas to look for, along with some of their better-known proprietary names. Also, two of the best U.S. importers for Spanish wine are Eric Solomon and Jorge Ordoñez.

BODEGAS BRETON

BODEGAS LAN

BODEGAS MONTECILLO—VIÑA CUMBRERO, VIÑA MONTY

BODEGAS MUGA—MUGA RESERVA, PRADO ENEA, TORREMUGA

BODEGAS REMIREZ DE GANUZIA

BODEGAS RIOJANAS—MONTE REAL, VIÑA ALBINA

C.U.N.E.—IMPERIAL, VIÑA REAL

CONTINO

FINCA ALLENDE

LA RIOJA ALTA—VIÑA ALBERDI, VIÑA ARDANZA

LOPEZ-HEREDIA

MARQUÉS DE CÁCERES

MARQUÉS DE MURRIETA

MARQUÉS DE RISCAL

MARTÍNEZ BUJANDA—CONDE DE VALDEMAR

PALACIOS REMONDO

REMELLURI

BEST BETS FOR RIOJA

1994* 1995* 2001** 2003 2004** 2005* 2006

Note: * signifies exceptional vintage   ** signifies extraordinary vintage

The two other famous winegrowing regions in Spain are Penedés (outside Barcelona) and the Ribera del Duero (between Madrid and Rioja). The most famous wine of the Penedés region is the sparkling wine called *cava,* of which the two best-known names in the United States are Codorniu and Freixenet. These are two of the biggest producers of bottle-fermented sparkling wine in the world, and one of the best features of these wines is their reasonable price. The Penedés region is also known for high-quality table wine. The major producer of this region (and synonymous with the quality of the area) is the

Torres family. Their famous wine, Gran Coronas Black Label, is made with 100 percent Cabernet Sauvignon, and is rare and expensive, though the Torres family produces a full range of fine Spanish wines in all price categories.

The Ribera del Duero has been around since the 1800s (though it was officially delimited in 1982), but is now becoming quite prominent in the United States. Some of the most expensive Spanish wines are from Ribera del Duero, such as Vega Sicilia. Other wines from that region, which are less expensive and more readily available, are Pesquera, Viña Mayor, Pago de los Capellanes, Bodegas Emilio Moro, Dominio de Pingus, Abadia Retuerta, Bodegas Aalto, Bodegas Felix Cullejo, Condado de Haza, and Haciendo Monasterio.

PRIORAT

This area just south of Barcelona near the Mediterranean produces some of the fullest-bodied wines of Spain, primarily from the Garnacha grape, sometimes blended with Syrah and Cabernet Sauvignon. Look for the wines of Alvaro Palacios, Mas Igneus, Costers del Siurana, Clas Erasmus, and Clos Martinet.

BEST BETS FOR RIBERA DEL DUERO

1996*   2001*   2003   2004**   2005*

*Note: * signifies exceptional vintage   ** signifies extraordinary vintage*

Have Spanish wines changed in style over the last twenty years?

Yes, without a doubt. As has been the case in countries like Chile and Italy, modern technology and new viticultural procedures have made for much better wines, many of which merit long-term aging. Hand selection of grapes, smaller barrels for fermentation, and using more French oak rather than the traditional American oak are all new changes in the wines of Spain.

There is also an increase in experimentation with Cabernet Sauvignon and single-vineyard bottlings.

BEST BETS FOR PENEDÉS AND PRIORAT

1998   1999   2000   2001*   2002*
2003*   2004*   2005*   2006

*Note: * signifies exceptional vintage*

FOR FURTHER READING

I recommend *The Wines of Rioja* by Hubrecht Duijker; *The New Spain* by John Radford; *The New and Classical Wines of Spain* by Jeremy Watson; and *Peñin Guide to Spanish Wine*.

The Wines of Australia

THERE ARE ABOUT FORTY DISTINCT winegrowing regions in Australia, with more than eighty districts and subdistricts. Do you need to know them all? Probably not, but to begin your Australian wine journey, you should be familiar with the best districts in four of Australia's six states:

NEW SOUTH WALES (N.S.W.)—Griffith, Hunter Valley, Mudgee

SOUTH AUSTRALIA (S.A.)—Barossa Valley, McLaren Vale, Coonawarra, Clare Valley

VICTORIA (VIC.)—Yarra Valley, Goulburn Valley, Rutherglen

WESTERN AUSTRALIA (W.A.)—Margaret River, Pemberton

The wine industry is by no means new to Australia. In fact, many of Australia's leading wine companies were established more than 175 years ago. Lindemans, Penfolds, Orlando, Henschke, and Seppelt are just a few of the companies that were founded during the nineteenth century. They are now among Australia's largest, or most prestigious, companies, and they continue to produce excellent wines.

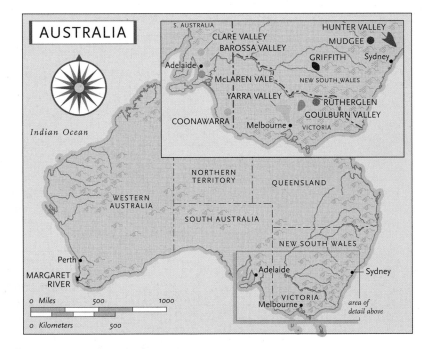

THERE ARE over 2,000 wineries in Australia.

AUSTRALIA IS now the second largest importer of wine to the United States. (Italy is the first.)

AMERICANS LIKE YELLOW TAIL

The number one imported wine in the United States is Yellow Tail. Sales rocketed from 200,000 cases in 2001 to almost 10 million cases in 2007.

TWENTY-TWO COMPANIES produce more than 80% of Australian wines.

THE AUSTRALIAN wine industry began in 1788 with the planting of Australia's first grapevines.

AUSTRALIAN WINE EXPORTS TO THE UNITED STATES (IN CASES)

1990	578,000
2007	20,000,000

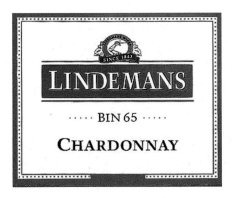

Which grape varieties are grown in Australia?

Major red-grape varieties are:

Shiraz: Called Syrah in the Rhône Valley of France and in California, it can produce spicy, robust, long-lived, and full-bodied wines. It is the most widely planted red grape in Australia at 101,597 acres (representing 25 percent of Australia's wine country).

Cabernet Sauvignon: As in Bordeaux and California, Australia's Cabernet Sauvignon grapes produce some of the best wines in the country. Always dry, Cabernet yields wines that range in style from medium to extremely full-bodied, depending on the producer and the region. It is often blended with Shiraz.

Pinot Noir: The great grape of Burgundy. In many cooler districts of Victoria, South Australia, and Western Australia, it is beginning to show signs of reaching quality levels similar to those of its famous brethren.

Merlot: A relative newcomer to Australia, but it is quickly gaining on Pinot Noir in terms of production.

The main white-grape varieties are:

Chardonnay: As in Burgundy and California, Australia's Chardonnay makes dry, full-flavored wines, as well as providing a base for sparkling wines. It is the most planted white grape in Australia.

Sémillon: In France, this grape is blended with Sauvignon Blanc to make white Bordeaux. In Australia it makes medium-style dry wines, is often blended with Chardonnay, and the best can benefit from aging. The Sémillon grape has been planted for more than two hundred years in Australia.

Riesling: A variety also grown in Germany; Alsace, France; Washington State; and New York State, this grape ranges in style from dry to sweet.

Sauvignon Blanc: Over the past few years, this grape has shown the largest increase in Australia's white-wine production.

What kinds of wine are produced in Australia?

Australia produces many different kinds of table wine, ranging from light, fruit-driven whites made from blends of grapes from different areas to vineyard-specific, barrel-aged reds from vines more than a hundred years old. It is not unusual to find Shiraz on original—not grafted—rootstock, because many Australian wine regions were not affected by phylloxera or other diseases.

Among the most interesting winemaking practices of Australia are the white blends made from Chardonnay and Sémillon and the red blends made

SOME OF the wine regions in Australia specialize in certain grapes. For example, Coonawarra is noted for its excellent Cabernet Sauvignon, Barossa for Shiraz, and the Hunter Valley for the white wines made from Sémillon.

IN THE 1830s, Cabernet Sauvignon vine cuttings from Château Haut-Brion in Bordeaux were planted near Melbourne. In 1832, James Busby brought back Shiraz cuttings from the Chapoutier vineyards in the Rhône Valley, which he planted in the Hunter Valley.

AUSSIE REDS ON THE MOVE

Only a few years ago, 65% of Australian wine grapes were white varieties. Today, white wines account for 45% and the reds contribute the remaining 55% to the harvest.

UP UNTIL the 1960s, Australian wine primarily consisted of sweet wine and portlike products. The Australians have graduated (with honors, I might add) to producing excellent dry table wines, but many of the best producers still pride themselves on luscious "stickies," the Australian term for dessert wine.

of Cabernet and Shiraz. Many of these represent high quality at reasonable prices, and they are also very enjoyable everyday wines.

How are wines labeled?

Effective with the 1990 vintage, the Australian wine industry's Label Integrity Program (LIP) took effect. Although it does not govern as many aspects as France's AOC laws, the LIP does regulate and oversee vintage, varietal, and geographical indication claims. To conform to LIP and other regulations set by the Australian Food Standards' Code, Australian wine labels, such as the one at left, give a great deal of information.

One of the most important pieces of information is the producer's name. In this case, the producer is Penfolds. If a single grape variety is printed on the label, the wine must be made from at least 85 percent of that variety. In a blend listing the varieties, as above, the percentages of each must be shown. (Note that the first grape listed is always the larger percentage.) If the label specifies a particular winegrowing district (for example, Clare Valley), at least 85 percent of the wine must originate there. If a vintage is given, 95 percent of the wine must be of that vintage.

What about vintages?

Australian winegrowing regions extend across three thousand miles with many diverse climatic and microclimatic conditions. As in California, there are not the extreme highs and lows that one may find in European vineyards, which is why I sometimes refer to them as "hassle-free" vintages. Case in point: Over the past ten years, Australia has had a string of good-to-great vintages.

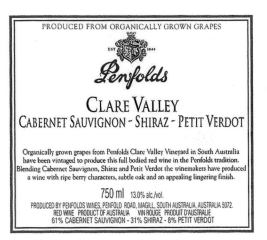

OF THE 150 different Australian wineries distributed in the United States, my list of the top wineries include: Hardy's, Lindemans, Orlando, Clarendon Hills, Michelton, Yalumba, Tahbilk, Rothbury, Penfolds, Petaluma, Rosemount, Wolf Blass, Henschke, Mountadam, D'Arenberg, Peter Lehmann, Leeuwin, Leasingham, Greg Norman Estate, Torbreck, Two Hands, and Mollydooker.

THE VINTAGE in Australia occurs in the first half of the year. The grapes are harvested from February to May. For any given vintage, Australia will have its wines approximately six months before Europe or America.

BIN NUMBERS are often found on Australian labels. These numbers were originally intended to indicate where the wines were stored in the cellar. Today they are used to denote the style or blend of a producer's wine.

BEST BETS FOR WINES FROM AUSTRALIA
(BAROSSA, MCLAREN, COONAWARRA)

1998*   2001**   2002   2003*   2004*   2005**   2006

*Note: * signifies exceptional vintage   ** signifies extraordinary vintage*

FOR FURTHER READING
I recommend *The Wine Atlas of Australia & New Zealand* by James Halliday, *James Halliday's Wine Companion*, *Crush* by Max Allen, and *Australian Wine: From the Vine to the Glass* by Patrick Iland and Peter Gago.

The Wines of Chile

Why is Chile the hottest wine-growing region in the world?

First and foremost is the quality of the wines for the price. In my opinion, the best value in red wines of the world comes from Chile. These are definitely the best South American wines today.

I did not include the wines of Chile in the first edition of this book, in 1985, but I followed the country's progress from the sidelines while tremendous expansion was taking place with plantings of international grape varieties. Since my most recent visit to wine country there, tasting the fruit of their labor, I am absolutely convinced that the Chileans are producing world-class wines, especially the reds.

Why did it take so long for Chile to produce world-class wines?

Chile has been making wine for more than 450 years, since the first grapes were planted in Chile in 1551 and the first wine was produced in 1555. Things really started getting interesting in the mid-1800s when French varietals such as Cabernet Sauvignon and Merlot were imported into Chile.

But all this progress came to a grinding halt in 1938 when the government of Chile decreed that no new vineyards could be planted. This law lasted until 1974. The renaissance of the modern wine industry of Chile really only began in the early 1980s, when the new technology of stainless-steel fermentors and the old technology of French oak barrels were combined to produce higher-quality wines.

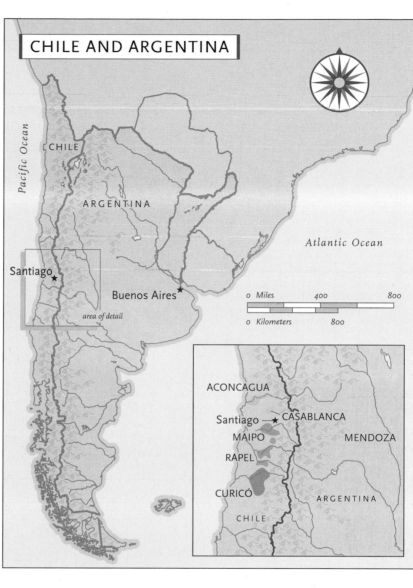

CHILE IS the fourth largest exporter of wine to the United States.

THE COUNTRY of Chile runs almost 2,600 miles north to south but only 60 miles wide at its broadest point.

THE NUMBER one Chilean wine imported into the United States is Concha y Toro. It is also one of the oldest wineries, started in 1883.

EXPORTS TO THE UNITED STATES

in 1996, Chile exported 100,000 cases of wine to the United States. Since then, Chilean wine exports have increased to more than 6 million cases.

IN 1995 THERE were only 12 wineries in Chile. Today there are more than 80.

WHICH GRAPE IS IT?

It was recently discovered that 40% of the Merlot planted in Chile is not really Merlot, but another grape called Carmenère.

IN 1985 THERE were only 30 acres of Chardonnay planted in Chile. In 2002 there were more than 15,000.

CABERNET SAUVIGNON accounts for 37% of the total acreage of premium grapes planted in Chile.

THE 1997 and 2003 vintages were two of the best ever produced in Chile.

CHILE IS a melting pot of Germans, French, Spanish, Italians, and Swiss.

What's happening in Chile today?

It's exciting. Going to Chile today is like going to California twenty-five years ago. And like California and Australia, there's no tradition of technique. Chile is a work in progress.

The Chilean winemakers have a lot going for them: The climate in Chile is somewhere between California and Bordeaux, and the Andes are spectacular. No matter where you are in the country, those majestic, snowcapped mountains always catch your eye.

What are the major winemaking regions in Chile?

There are twelve different regions for wine production in Chile. The seven most important are:

ACONCAGUA VALLEY   MAIPO VALLEY
CASABLANCA VALLEY   MAULE
COLCHAGUA VALLEY   RAPEL VALLEY
CURICÓ VALLEY

What are the major grapes grown in Chile?

The main white varieties are:

SAUVIGNON BLANC
CHARDONNAY
MUSCAT

The main red varieties are:

CABERNET SAUVIGNON   PAÍS (MISSION GRAPE)
CARMENÈRE   SYRAH
MERLOT

PRE-PHYLLOXERA

To this day, Chile can still boast about having pre-phylloxera wines. As you may recall, the phylloxera louse destroyed all the vineyards in France in the 1870s and the Chileans brought over their vine cuttings prior to this infestation. To this day, the phylloxera bug has not been a problem in Chile.

How do I buy the Chilean wines?

Almost all Chilean wines are dry, and there are no official classifications. Most of the wineries in Chile produce wines of different quality levels and price points, and label their wines by grape variety. The wine laws are very much like those in California. For example, if the grape variety is named on the label, the wine must contain a minimum of 85 percent of that grape. The best Chilean wineries also have their own special names for their highest-quality wines.

The best way to buy Chilean wines is from the top producers. The following is a list of my favorite wineries. In parentheses are names of the best wines:

CALITERRA (RESERVE)
CARMEN (VIDURE CABERNET RESERVE)
CASA LAPOSTOLLE (CUVÉE ALEXANDRE, CLOS APALTA)
CONCHA Y TORO (DON MELCHOR, ALMAVIVA)
COUSIÑO MACUL (FINIS TERRAE, ANTIGUAS RESERVAS)
DOMAINE PAUL BRUNO
VIÑA ERRAZÚRIZ (DON MAXIMIANO)
LOS VASCOS (RESERVA DE FAMILIA)
VIÑA MONTES (ALPHA M)
SEÑA
UNDURRAGA (RESERVE)
VERAMONTE (PRIMUS)
VIÑA SANTA CAROLINA
VIÑA SANTA RITA (CASA REAL)
VIÑEDO CHADWICK
VIÑA ALMAVIVA

Will Chilean wines age well?

On my most recent visit to Chile, I was lucky enough to taste the 1969, 1965, and 1960 vintages of Cousiño Macul Antiguas Cabernet Sauvignon, which were just coming into their prime.

BEST BETS FOR CHILE: RED VINTAGES

1997*   1999*   2001*   2002
2003*   2004   2005*

Note: * signifies exceptional vintage

THE TOP FIVE CHILEAN WINE BRANDS IN THE UNITED STATES

1. Concha y Toro
2. Walnut Crest
3. San Pedro
4. Santa Rita
5. Santa Carolina

Source: Impact Databank

WINE REGIONS AND THEIR SPECIALTIES

Region	Specialty
Casablanca	Chardonnay, Sauvignon Blanc
Maipo	Cabernet Sauvignon
Rapel	Merlot, Carmenère

The Best of Chile
Almaviva
Montes (Alpha M)
Seña
Le Dix de Los Vascos
Errazúriz (Don Maximiano)
Cousiño-Macul (Antiguas Reservas)

**THE SEVEN LARGEST WINERIES
IN CHILE**

Winery	Date Founded
Concha y Toro	1883
San Pedro	1865
Santa Rita	1880
Santa Carolina	1875
Errazúriz	1870
Undurraga	1885
Viña Canepa	1930

THE FRENCH CONNECTION
(FRENCH INVESTMENT IN CHILE)

FRENCH INVESTOR	CHILEAN WINERY
Grand Marnier	Casa Lapostolle
Domaines Lafite Rothschild	Los Vascos
Bruno Prats & Paul Pontalier	Aquitania/Paul Bruno
William Fèvre	Fèvre
Château Larose-Trintaudon	Casas del Toqui
Baron Philippe de Rothschild	Concha y Toro
	(Viña Almaviva Puente Alto)

OTHER FOREIGN INVESTORS IN CHILE INCLUDE:

Quintessa, California	Veramonte
Torres Winery, Spain	Miguel Torres Winery
Kendall-Jackson, California	Viña Calina

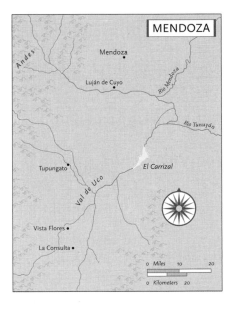

MENDOZA

The Wines of Argentina

What's happening in Argentina?

Great things are happening in Argentina's wine industry! A growing number of wineries have changed their philosophy and now concentrate on producing higher-quality wines rather than manufacturing large quantities.

Until recently, Argentina was more interested in producing inexpensive bulk wines. Now its winemakers are beginning to understand the worldwide demand for quality wines. Still, today 90 percent of Argentina's wine is consumed in Argentina, which is the problem.

As it was in Spain and Italy twenty-five years ago (and Chile twenty years ago), wineries were confronted with the dilemma of whether to produce wine for the domestic market—wine that was sometimes bland and oxidized—or try to make better wine for the export market. Today a new, higher style of Argentinean wine is emerging.

Why has it taken so long for Argentina to become a player in the wine world?

In the 1970s and '80s, the economic crisis in Argentina prevented winemakers from producing great wine. Any wine region must have capital to produce great wine. In fact, Argentina's wine industry is booming because more investment has been poured into it in the last ten years than in the previous fifty!

Who's investing in Argentina?

From California, Jess Jackson (Kendall-Jackson) and Paul Hobbs; from Bordeaux, the Lurton family and also the owners of Château Cheval Blanc, who have created a wine called Cheval des Andes; and the Domaines Barons de Rothschild (Lafite). Even wineries from Chile, such as Concha y Toro and Santa Rita, are investing their money and expertise in Argentina.

Why is Argentina easier to understand than Chile?

Think red. Think Mendoza. Think Malbec. The best wines of Argentina are red. The major region is Mendoza, which produces 75 percent of all wines coming from Argentina. Malbec, with 43,000 acres, is the best-quality grape planted in Mendoza. Other red grapes produced are Cabernet Sauvignon, with 30,000 acres, and Merlot. In the future, look for Syrah. Argentina is also producing world-class Chardonnay.

How do I buy wine from Argentina?

As with Chile, buy from the best producers. There are more than 1,000 wineries in Argentina. Here is a list of some of the best:

ACHÁVAL-FERRER	J&F LURTON	PAUL HOBBS
BODEGAS ESMERALDA	LA RURAL	SALENTEIN
CATENA ZAPATA	LOPEZ	TIKA
CHEVAL DES ANDES	LUCA	TRAPICHE
ETCHART	LUIS BOSCA	VALENTIN BIANCHI
FINCA FLICHMAN	MARIPOSA	WEINERT
NORTON		

ARGENTINE WINE exports to the U.S. has tripled since 2002.

WITH 300 days of sunshine and only eight inches of rain annually, the Argentineans have set up an elaborate network of canals and dams to irrigate their vineyards.

ARGENTINA HAS been producing wine for more than 400 years. The first grapes in Argentina were planted in 1554.

ARGENTINA IS the largest producer of wines in South America. It's the fifth largest wine producer in the world. Argentina is also the sixth largest consumer of wines in the world.

OVER 70% of the 500,000 acres of vineyards planted in Argentina are in Mendoza—about 360,000 acres.

ALL ARGENTINEAN varietal wines are 100% of the grape named on the label.

BEST BETS FOR WINES FROM ARGENTINA

2002*	2003*	2004*	2005*	2006	2007

Note: * signifies exceptional vintage

Questions for Class Seven: Wines of the World: Italy, Spain, Australia, Chile, and Argentina

AUSTRALIA

CHILE

    a. Clos Apalta   ____Santa Rita
    b. Don Melchor   ____Veramonte
    c. Antiguas Reservas   ____Casa Lapostolle
    d. Don Maximiano   ____Viña Montes
    e. Alpha M   ____Santa Rita
    f. Casa Real   ____Concha y Toro
    g. Primus   ____Cousiño Macul

ARGENTINA

Champagne, Sherry, and Port

CHAMPAGNE · *MÉTHODE CHAMPENOISE* · STYLES OF CHAMPAGNE ·

OPENING CHAMPAGNE · CHAMPAGNE GLASSES ·

SPARKLING WINE · SHERRY · PORT

THE CHAMPAGNE region covers has
43,680 acres of vineyards.

THERE ARE OVER 20,000 growers who mostly
sell to the 250 négociants (shippers).

Champagne and Fortified Wines

NOW WE'RE BEGINNING our last class—the last chapter on the wine itself. This is where the course ends—on a happy note, I might add. What better way to celebrate than with Champagne?

Why do I group Champagne, Sherry, and Port together? Because as diverse as these wines are, the way the consumer will buy them is through the reputation and reliability of the shipper. Since these are all blended wines, the shipper is responsible for all phases of the production—you concern yourself with the house style. In Champagne, for example, Moët & Chandon is a well-known house; in Port, the house of Sandeman; and in Sherry, the house of Pedro Domecq.

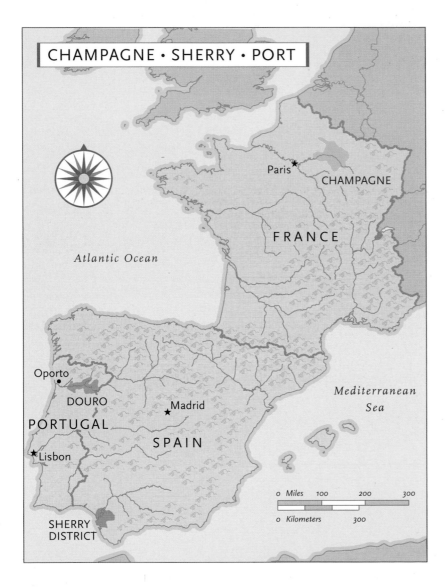

CHAMPAGNE

What's Champagne?

Everyone knows that Champagne is a sparkling bubbly that everyone drinks on New Year's Eve. It's more than that. Champagne is a region in France—the country's northernmost winemaking region, to be exact—and it's an hour and a half northeast of Paris.

Why do I stress its northern location? Because this affects the taste of the wines. In the Champagne region, the grapes are picked with higher acidity than in most other regions, which is one of the reasons for Champagne's distinct taste. The Champagne region is divided into four main areas:

VALLEY OF THE MARNE

MOUNTAIN OF REIMS

CÔTE DES BLANCS

CÔTE DES BAR

Three grapes can be used to produce Champagne:

Pinot Noir: Accounts for 38 percent of all grapes planted
Pinot Meunier: Accounts for 35 percent of all grapes planted
Chardonnay: Accounts for 27 percent of all grapes planted

In France, only sparkling wines that come from the region of Champagne may be called "Champagne." Some American producers have borrowed the name *Champagne* to put on the label of their sparkling wines. These cannot and should not be compared with Champagne from France.

What are the three major types of Champagne?

Non-vintage/multiple vintage: A blend of two or more harvests, 60–80% base wine from current harvest and 20–40% wine from previous vintages.
Vintage: From a single vintage
"Prestige" cuvée: From a single vintage with longer aging requirements

Why is there such a tremendous price difference between non-vintage and "prestige" cuvée Champagnes?

"Prestige" Champagnes usually meet the following requirements to be designated as such:

- Made from the best grapes of the highest-rated villages.
- Made from the first pressing of the grapes.
- Spent more time aging in the bottle than non-vintage Champagnes.
- Made only in vintage years.
- Made in small quantity, and the demand is high. Price is dictated largely by supply and demand.

Acidity in Champagne not only gives freshness to the wines, but it is also important to their longevity, and it stimulates the palate before lunch or dinner.

THE BALANCE of the fruit and acidity, together with the bubbles (CO_2), are what make good Champagne.

NON-VINTAGE Champagne is more typical of the house style than vintage Champagne.

VINTAGE CHAMPAGNE must contain 100% of that vintage year's harvest.

MORE THAN 80% of the Champagnes produced are not vintage dated. This means they are blends of several years' wines.

SHIPPERS DON'T always agree on the quality of the wines produced in any given vintage, so the years for vintage Champagnes vary from shipper to shipper. Historically, each house usually declares a vintage three years out of each decade.

TWO METHODS of making rosé Champagne: 1) add red wine to the blend; 2) leave the red grape skins in contact with the must for a short period of time.

MOST CHAMPAGNES are fermented in stainless steel.

CLASSIC CHAMPAGNES (non-vintage) must be aged for a minimum of 15 months in the bottle after bottling. Vintage Champagnes must be aged for a minimum of three years after bottling.

DOM PÉRIGNON Champagne is aged six to eight years before it is put on the market.

"My dear girl, there are some things that just aren't done, such as drinking Dom Perignon '53 above the temperature of 38 degrees Fahrenheit. That's just as bad as listening to the Beatles without earmuffs!"
—SEAN CONNERY *as* JAMES BOND, *in Goldfinger (1964)*

Is every year a vintage year?

No, but more recently, 1995*, 1996*, 1998, 1999, 2000, and 2002* were. Note: These were vintage years for most Champagne houses. "Vintage" in Champagne is different from other wine regions, because each house makes its own determination on whether or not to declare a vintage year.

Note: * signifies exceptional vintage

How is Champagne made?

Champagne is made by a process called *Méthode Champenoise*. When a similar method is used outside La Champagne, it is called *Méthode Traditionnelle* or Classic Method or *Método Tradicional*, etc. The use of the expression *Méthode Champenoise* is not allowed in the European Union outside of Champagne.

MÉTHODE CHAMPENOISE

Harvest—The normal harvest usually takes place in late September or early October.

Pressing the Grapes—Only two pressings of the grapes are permitted. Prestige cuvée Champagnes are usually made exclusively from the first pressing. The second pressing, called the *taille*, is generally blended with the cuvée to make vintage and non-vintage Champagnes.

Fermentation—All Champagnes undergo a first fermentation when the grape juice is converted into wine. Remember the formula: Sugar + Yeast = Alcohol + CO_2. The carbon dioxide dissipates. The first fermentation takes two to three weeks and produces still wines.

Blending—The most important step in Champagne production is the blending of the still wines. Each of these still wines is made from a single grape variety from a single village of origin. The winemaker has to make many decisions here. Three of the most important ones are:

1. Which grapes to blend—how much Chardonnay, Pinot Noir, and Pinot Meunier?

2. From which vineyards should the grapes come?

3. Which years or vintages should be blended? Should the blend be made from only the wines of the harvest, or should several vintages be blended together?

Liqueur de Tirage—After the blending process, the winemaker adds *Liqueur de Tirage* (a blend of sugar and yeast), which will begin the wine's second fermentation. At this point, the wine is placed in its permanent bottle with a temporary bottle cap.

Second Fermentation—During this fermentation, the carbon dioxide stays in the bottle. This is where the bubbles come from. The second fermentation also leaves natural sediments in the bottle. Now the problems begin. How do you get rid of the sediments without losing the carbon dioxide? Go on to the next steps.

Aging—The amount of time the wine spends aging on its sediments is one of the most important factors in determining the quality of the wine.

Riddling—The wine bottles are now placed in A-frame racks, necks down. The *remueur*, or riddler, goes through the racks of Champagne bottles and gives each bottle a slight turn while gradually tipping the bottle farther downward. After six to eight weeks, the bottle stands almost completely upside down, with the sediments resting in the neck of the bottle.

Dégorgement—The top of the bottle is dipped into a brine solution to freeze it, and then the temporary bottle cap is removed and out fly the frozen sediments, propelled by the carbon dioxide.

Dosage—A combination of wine and cane sugar is added to the bottle after *dégorgement*. At this point, the winemaker can determine whether he wants a sweeter or a drier Champagne.

Recorking—The wine is recorked with real cork instead of a bottle cap.

DOSAGE

The dosage determines whether the wine will be dry, sweet, or any style in between. The following shows you the guidelines the winemaker uses when he adds the dosage.

Brut: *Dry*
Extra dry: *Semidry*
Sec: *Semisweet*
Demi-sec: *Sweet*

COMMON CHAMPAGNE AROMAS:

Apple	Yeast (bread dough)
Toast	Hazelnuts/walnuts
Citrus	

WOMEN AND CHAMPAGNE

Women, particularly ones attached to royal courts, deserve much of the credit for Champagne's international fame. Madame de Pompadour said that Champagne was the only drink that left a woman still beautiful after drinking it. Madame de Parabère once said that Champagne was the only wine to give brilliance to the eyes without flushing the face.

IT IS RUMORED that Marilyn Monroe once took a bath in 350 bottles of Champagne. Her biographer George Barris said that she drank and breathed Champagne "as if it were oxygen."

WHEN A London reporter asked Madame Lilly Bollinger when she drank Champagne, Madame Bollinger replied: "I drink it when I'm happy and when I'm sad. Sometimes I drink it when I'm alone. When I have company I consider it obligatory. I trifle with it if I'm not hungry and drink it when I am. Otherwise I never touch it—unless I'm thirsty."

UNTIL AROUND 1850, all Champagne was sweet.

OCCASIONALLY A Champagne will be labeled "extra brut," which is drier still than brut.

BRUT AND extra-dry are the wines to serve as apéritifs, or throughout the meal. Sec and demi-sec are the wines to serve with desserts and wedding cake!

BLANC DE BLANCS Champagne is made from 100% Chardonnay.

BLANC DE NOIR Champagne is made from 100% Pinot Noir.

THERE ARE more than approximately 4,000 Champagne producers.

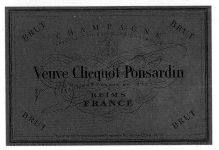

What accounts for the different styles of Champagne?

Going back to the three grapes we talked about that are used to make Champagne, the general rule is: The more white grapes in the blend, the lighter the style of the Champagne. And the more red grapes in the blend, the fuller the style of the Champagne.

Also, some producers ferment their wines in wood. Bollinger ferments some, and Krug ferments all their wines this way. This gives the Champagne fuller body and bouquet than those fermented in stainless steel.

How do I buy a good Champagne?

First, determine the style you prefer, whether full-bodied or light-bodied, a dry brut or a sweet demi-sec. Then make sure you buy your Champagne from a reliable shipper/producer. Each producer takes pride in its distinctive house style, and strives for a consistent blend, year after year. The following are some brands in national distribution to look for. While it is difficult to be precise, the designations generally conform to the style of the houses.

LIGHT, DELICATE

A. Charbaut et Fils
Jacquesson
Lanson

LIGHT TO MEDIUM

Billecart-Salmon
Deutz
Nicolas Feuillatte
Laurent-Perrier
G.H. Mumm
Perrier-Jouët
Pommery
Ruinart Père & Fils
Taittinger

MEDIUM

Charles Heidsieck
Moët & Chandon
Piper-Heidsieck
Pol Roger
Salon

MEDIUM TO FULL

Henriot
Louis Roederer

FULL, RICH

Bollinger
A. Gratien
Krug
Veuve Clicquot

When is Champagne ready to drink?

As soon as you buy it. Champagne is something you can drink right away. Non-vintage Champagnes are meant to be drunk within two to three years, and vintage and prestige cuvée Champagnes can be kept longer, about ten to fifteen years. So if you're still saving that Dom Pérignon that you received for your tenth wedding anniversary fifteen years ago, don't wait any longer. Open it!

What's the correct way to open a bottle of Champagne?

Before we sip Champagne in class, I always take a few moments to show everyone how to open a bottle of Champagne properly. I do this for a good reason. Opening a bottle of Champagne can be dangerous, and I'm not kidding. If you know the pounds per square inch that are under pressure in the bottle, you know what I'm talking about.

OPENING CHAMPAGNE CORRECTLY

1. It is especially important that the bottle be well chilled before you open it.

2. Cut the foil around the top of the bottle.

3. Place your hand on top of the cork, never removing your hand until the cork is pulled out completely. (I know this may seem a bit awkward, but it's very important.)

4. Undo the wire. Either leave it on the cork or take it off carefully.

5. Carefully put a cloth napkin over the top of the cork; if the cork pops, it will go safely into the napkin.

6. Remove the cork gently, slowly turning the bottle in one direction and the cork in another. The idea behind opening a bottle is to ease the cork out gently rather than cracking the bottle open with a loud pop and letting it foam. That may be a lot of fun, but it does nothing for the Champagne. When you pop off the cork, you allow the carbon dioxide to escape. That carbon dioxide is what gives Champagne its sparkle. If you open a bottle of Champagne in the way I've just described, it can be opened hours before your guests arrive with no loss of carbon dioxide.

"It's not a Burgundy; it's not a Bordeaux; it's a white wine; it's a sparkling wine that should be kept no longer than two to three years. It should be consumed young."
—CLAUDE TAITTINGER

THE TOP FIVE CHAMPAGNE HOUSES IN SHIPMENTS TO THE UNITED STATES IN 2007

1. Moët & Chandon
2. Veuve Clicquot
3. Perrier-Jouët
4. Piper Heidsieck
5. Nicolas Feuillatte

CHAMPAGNE HOUSES market about two-thirds of Champagne's wines, but they own less than 10% of the vineyards.

THE PRESSURE in a bottle of Champagne is close to 90 pounds per square inch (or "six atmospheres," or roughly three times the pressure in your automobile tire). Champagne is put into heavy bottles to hold the pressurized wine. This is another reason why Champagne is more expensive than ordinary wine.

MILLENNIUM MADNESS

In 1999, a record 327 million bottles of Champagne were sold.

Which glasses should Champagne be served in?

AS BEAUTIFUL as Helen was, the resulting glass was admittedly wide and shallow.

TO EVALUATE Champagne, look at the bubbles. The better wines have smaller bubbles and more of them. Also, with a good Champagne, the bubbles last longer. Bubbles are an integral part of the wines of Champagne. They create texture and mouth feel.

HOW MANY bubbles are in a bottle of Champagne? According to scientist Bill Lembeck, 49 million per bottle!

CHAMPAGNE BOTTLE SIZES

Magnum	2 bottles
Jeroboam	4 bottles
Rehoboam	6 bottles
Methuselah	8 bottles
Salmanazar	12 bottles
Balthazar	16 bottles
Nebuchadnezzar	20 bottles

No matter which Champagne you decide to serve, you should serve it in the proper glass. There's a little story behind the Champagne glass, dating back to Greek mythology. The first *coupe*, a footed glass with a shallow cup that widens toward the rim, was said to be molded from the breast of Helen of Troy. The Greeks believed that wine drinking was a sensual experience, and it was only fitting that the most beautiful woman take part in shaping the chalice.

Centuries later, Marie Antoinette, Queen of France, decided it was time to create a new Champagne glass. She had coupes molded to her own breasts, which changed the shape of the glass entirely, since Marie Antoinette was, shall we say, a bit more well endowed than Helen of Troy.

The glasses shown to the left are the ones commonly used today—the flute and the tulip-shaped glass. Champagne does not lose its bubbles as

WINE AND FOOD

Champagne is one of the most versatile wines that you can drink with a number of foods, from apéritif to dessert. Here are some Champagne-and-food combinations that experts suggest:

CLAUDE TAITTINGER: *Mr. Taittinger's general rule is: "Never with sweets." Instead, he suggests "a Comtes de Champagne Blanc de Blancs drunk with seafood, caviar, or pâté of pheasant." Another note from Mr. Taittinger: He doesn't serve Champagne with cheese because, he says, "The bubbles do not go well." He prefers red wine with cheese.*

CHRISTIAN POL ROGER: *With Brut non-vintage: light hors d'oeuvres, mousse of pike. With vintage: pheasant, lobster, other seafood. With rosé: a strawberry dessert.*

quickly in these glasses as it did in the old-fashioned model, and these shapes also enhance the smell and aromas of the wine in the glass.

What's the difference between Champagne and sparkling wine?

As I've already mentioned, Champagne is the wine that comes from the Champagne region of France. In my opinion, it is the best sparkling wine in the world, because the region has the ideal combination of elements conducive to excellent sparkling winemaking. The soil is fine chalk, the grapes are the best grown anywhere for sparkling wine, and the location is perfect. This combination of soil, climate, and grapes is reflected in the wine.

Sparkling wine, on the other hand, is produced in many areas, and the quality varies from wine to wine. The Spanish produce the popular Codorniu and Freixenet—both excellent values and good sparkling wines, known as *cavas.* The German version is called *Sekt.* Italy has *spumante,* which means "sparkling." The most popular Italian sparkling wine in the United States is Asti Spumante.

New York State and California are the two main producers of sparkling wine in this country. New York is known for Great Western, Taylor, and Gold Seal. California produces many fine sparkling wines, such as Domaine Chandon, Korbel, Piper-Sonoma, Schramsberg, Mumm Cuvée Napa, Roederer Estate, Domaine Carneros, Iron Horse, Scharffenberger, and "J," by Jordan Winery. Many of the larger California wineries also market their own sparkling wines.

DOMAINE CHANDON is owned by the Moët-Hennessy Group, which is responsible for the production of Dom Pérignon in France. In fact, the same winemaker is flown into California to help make the blend for the Domaine Chandon.

ABOUT 20% OF the sparkling wines made in the United States are made by the Méthode Champenoise.

PIPER-HEIDSIECK has sold the Piper-Sonoma and its vineyards to Jordan Winery. Jordan will now produce the wines for Piper-Sonoma.

Is there a difference between the way Champagne and sparkling wines are made?

Sometimes. All authentic Champagnes and many fine sparkling wines are produced by *Méthode Champenoise,* described earlier in this chapter, and which, as you now know, is laborious, intensive, and very expensive. If you see a bottle of sparkling wine for $3.99, you can bet that the wine was not made by this process. The inexpensive sparkling wines are made by other methods. For example, in one method the secondary fermentation takes place in large tanks. Sometimes these tanks are big enough to produce 100,000 bottles of sparkling wine.

SHERRY

The two greatest fortified wines in the world are Port and Sherry. These wines have much in common, although the end result is two very different styles.

What exactly is fortified wine?

Fortified wine is made when a neutral grape brandy is added to wine to raise the wine's alcohol content. What sets Port apart from Sherry is when the winemaker adds the neutral brandy. It's added to Port during fermentation. The extra alcohol kills that yeast and stops the fermentation, which is why Port is relatively sweet. For Sherry, on the other hand, the brandy is added after fermentation.

Where is Sherry made?

Sherry is produced in sunny southwestern Spain, in Andalusia. An area within three towns makes up the Sherry triangle. They are:

> **Jerez de la Frontera**
>
> **Puerto de Santa María**
>
> **Sanlúcar de Barrameda**

Which grapes are used to make Sherry?

There are two main varieties:

> *Palomino* (this shouldn't be too difficult for horse lovers to remember)
> *Pedro Ximénez* (named after Peter Siemons, who brought the grape from Germany to Sherry)

What are the different types of Sherry?

> *Manzanilla:* Dry
> *Fino:* Dry
> *Amontillado:* Dry to medium-dry
> *Oloroso:* Dry to medium-dry
> *Cream:* Sweet

ANOTHER FORTIFIED wine is Madeira. Although it is not as popular as it once was, Madeira wine was probably the first wine imported into America. It was favored by the colonists, including George Washington, and was served to toast the Declaration of Independence.

TWO OTHER famous fortified wines are Marsala (from Italy) and Vermouth (from Italy and France).

THE NEUTRAL grape brandy, when added to the wine, raises the alcohol content to 15% to 20%.

FOR YOU HISTORIANS, Puerto de Santa María is where Christopher Columbus's ships were built and where all the arrangements were made with Queen Isabella for his journey of discovery.

THE PALOMINO grape accounts for 90% of the planted vineyards in Sherry.

What are the unique processes that characterize Sherry production?

Controlled oxidation and fractional blending. Normally a winemaker guards against letting any air into the wine during the winemaking process. But that's exactly what makes Sherry—the air that oxidizes the wine. The winemaker places the wine in barrels and stores it in a bodega.

What's a bodega?

No, I'm not talking about a Latino grocery store at 125th Street and Lexington Avenue in New York City. In Sherry, a *bodega* is an aboveground structure used to store wine. Why do you think winemakers would want to store the wine above ground? For the air. Sherry is an oxidized wine. They fill the barrels approximately two-thirds full, instead of all the way, and they leave the bung (cork) loosely in the barrel to let the air in.

THE ANGEL'S SHARE

When Sherry is made, winemakers let air into the barrels, and some wine evaporates in the process. Each year they lose a minimum of 3 percent of their Sherry to the angels, which translates into thousands of bottles lost through evaporation!

Why do you think the people of Sherry are so happy all the time? Besides the excellent sunshine they have, the people breathe in oxygen and Sherry.

So much for controlled oxidation. Now for fractional blending. Fractional blending is carried out through the Solera System.

What's the Solera System?

The Solera System is an aging and maturing process that takes place through the dynamic and continuous blending of several vintages of Sherry that are stored in rows of barrels. At bottling time, wine is drawn out of these barrels— never more than one-third the content of the barrel—to make room for the new vintage. The purpose of this type of blending is to maintain the "house" style of the Sherry by using the "mother" wine as a base and refreshing it with a portion of the younger wines.

HERE'S ANOTHER abbreviation for you— PX. Do you remember TBA, QbA, AOC, and DOC? If you want to know Sherry, you may have to say "PX," which stands for the Pedro Ximénez grape.

I'M SURE you're familiar with these four top-selling Sherries: Harveys Bristol Cream, Dry Sack, Tio Pepe, and La Ina.

PX IS USED to make Cream Sherry, like Harveys Bristol Cream, among others. Cream Sherry is a blend of PX and Oloroso.

SHERRY ACCOUNTS for less than 3% of Spanish wine production.

IN TODAY'S Sherry, only American oak is used to age the wine.

SOME SOLERAS can be a blend of 10 to 20 different harvests.

TO CLARIFY the Sherry and rid it of all sediment, beaten egg whites are added to the wine. The sediment attaches itself to the egg whitesand drops to the bottom of the barrel. The question always comes up: "What do they do with the yolks?" Did you ever hear of flan? That's the puddinglike dessert made from all the yolks. In Sherry country, this dessert is called *tocino de cielo*, which translated means "the fat of the angels."

OF THE SHERRY consumed in Spain, 90% is Fino and Manzanilla. As one winemaker said, "We ship the sweet and drink the dry."

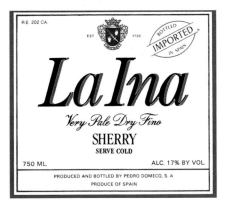

How do I buy Sherry?

Your best guide is the producer. It's the producer, after all, who buys the grapes and does the blending. Ten producers account for 60 percent of the export market. The top Sherry producers are:

CROFT	OSBORNE
EMILIO LUSTAU	PEDRO DOMECQ
GONZÁLEZ BYASS	SANDEMAN
HARVEYS	SAVORY AND JAMES
HIDALGO	WILLIAMS & HUMBERT

How long does a bottle of Sherry last once it's been opened?

Sherry will last longer than a regular table wine, because of its higher alcoholic content, which acts as a preservative. But once Sherry is opened, it will begin to lose its freshness. To drink Sherry at its best, you should consume the bottle within two weeks of opening it and keep the opened bottle refrigerated. Manzanilla and Fino Sherry should be treated as white wines and consumed within a day or two.

FOR FURTHER READING
I recommend *Sherry* by Julian Jeffs.

WINE AND FOOD

MAURICIO GONZÁLEZ: *He believes that Fino should always be served well chilled. He enjoys having Fino as an apéritif with Spanish tapas (hors d'oeuvres), but he also likes to complement practically any fish meal with the wine. Some of his suggestions: clams, shellfish, lobster, prawns, langoustines, fish soup, or a light fish such as salmon.*

JOSÉ IGNACIO DOMECQ: *He suggests that very old and rare Sherry should be served with cheese. Fino and Manzanilla can be served as an apéritif or with light grilled or fried fish, or even smoked salmon. "You get the taste of the* smoke *better than if you have it with a white wine." Amontillado is not to be consumed like a Fino. It should be served with light cheese, chorizo (sausage), ham, or shish kebab. It is a perfect complement to turtle soup or a consommé. According to Mr. Domecq, dry Oloroso is known as a sporty drink in Spain—something to drink before hunting, riding, or sailing on a chilly morning. With Cream Sherry, Mr. Domecq recommends cookies, pastries, and cakes. Pedro Ximénez, however, is better as a topping for vanilla ice cream or a dessert wine before coffee and brandy.*

PORT

Port comes from the Douro region in northern Portugal. In fact, in recent years, to avoid the misuse of the name "Port" in other countries, the true Port wine from Portugal has been renamed "Porto" (for the name of Oporto, the port city from which it's shipped).

Just a reminder: Neutral grape brandy is added to Port during fermentation, which stops the fermentation and leaves behind up to 9 to 11 percent residual sugar. This is why Port is on the sweet side.

What are the two types of Port?

Cask-aged Port: This includes *Ruby Port*, which is dark and fruity, blended from young non-vintage wines (Cost: $); *Tawny Port*, which is lighter and more delicate, blended from many vintages (Cost: $$); *Aged Tawny*, which is aged in casks—sometimes up to forty years and longer (Cost: $$/$$$); and *Colheita*, which is from a single vintage but wood-aged a minimum of seven years (Cost: $$/$$$$).

Bottle-aged Port: These wines include *Late Bottled Vintage (LBV)*, which is made from a single vintage, bottled four to six years after the harvest, and similar in style to vintage Port, but lighter, ready to drink on release, with no decanting needed (Cost: $$$); *Vintage Character*, which is similar in style to LBV, but is made from a blend of vintages from the better years (Cost: $$); *Quinta*, which is from a single vineyard (Cost: $$$/$$$$); and *Vintage Port*, which is aged two years in wood and will mature in the bottle over time (Cost: $$$$).

PORT WINE has been shipped to England since the 1670s. During the 1800s, to help preserve the Port for the long trip, shippers fortified it with brandy, resulting in Port as we know it today.

AS WITH SHERRY, Port evaporation is a problem—some 15,000 bottles evaporate into the air every year.

PORT IS usually 20% alcohol. Sherry, by comparison, is usually around 18%.

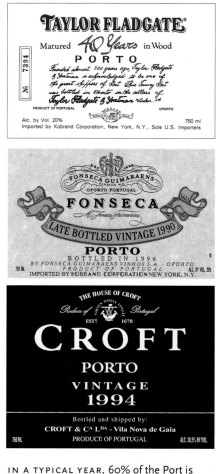

IN A TYPICAL YEAR, 60% of the Port is Tawny and Ruby; 30% is Vintage Character; 7% is Aged Tawny; and 3% is Vintage Port.

WOOD PORT VS. VINTAGE PORT

The biggest difference between cask-aged Port (such as Ruby and Tawny) and bottle-aged Port is this: The cask-aged Port is ready to drink as soon as it is bottled and it will not improve with age. Bottle-aged Port, on the other hand, gets better as it matures in the bottle. A great vintage Port will be ready to drink fifteen to thirty years after the vintage date, depending upon the quality of the vintage.

THE FIRST vintage Port was recorded in 1765.

How do I buy Port?

Once again, as with Sherry, Port's grape variety should not dictate your choice. Find the style and the blend you prefer, but even more important, look for the most reliable producers. Of the Port available in the United States, the most important producers are:

A. A. FERREIRA	NIEPOORT & CO., LTD.
C. DA SILVA	QUINTA DO NOVAL
CHURCHILL	RAMOS PINTO
COCKBURN	ROBERTSON'S
CROFT	SANDEMAN
DOW	TAYLOR FLADGATE
FONSECA	W. & J. GRAHAM
HARVEYS OF BRISTOL	WARRE & CO.

BEST BETS FOR VINTAGES OF PORT

1963*   1970*   1977*   1983*   1985   1991*
    1992   1994*   1997*   2000*   2003*

Note: * signifies exceptional vintage

THE BRITISH are known to be Port lovers. Traditionally, upon the birth of a child, parents buy bottles of Port to put away for the baby until its 21st birthday, not only the age of maturity of a child, but also that of a fine Port.

"The 1994 vintage is the greatest for Port since the legendary 1945."
—JAMES SUCKLING, *Vintage Port*

Should vintage Port be decanted?

Yes, because you are likely to find sediment in the bottle. By making it a practice to decant vintage Port, your enjoyment of it will be enhanced.

How long will Port last once it's been opened?

Port has a tendency to last longer than ordinary table wine because of its higher alcohol content. But if you want to drink Port at its prime, drink the contents of the open bottle within one week.

FOR FURTHER READING

I recommend *The Port Companion* by Godfrey Spence and *Vintage Port: The Wine Spectator's Ultimate Guide* by James Suckling.

Is every year a vintage year for Port?

No, it varies from shipper to shipper. And in some years, no vintage Port is made at all. For example, in 1994, 1997, 2000, and 2003, four of the recent vintages for Port, most producers declared a vintage. On the other hand, in 1990 and 1993, Port, in general, was not considered vintage quality.

Questions for Class Eight:
Champagne, Sherry, and Port

The Greater World
of Wine

❧

THE COMPLETE WINE-TASTING COURSE • THE PHYSIOLOGY OF TASTING

WINE • MATCHING WINE AND FOOD • WINE-BUYING STRATEGIES

• CREATING AN EXEMPLARY RESTAURANT WINE LIST

• WINE SERVICE AND WINE STORAGE

• FREQUENTLY ASKED QUESTIONS ABOUT WINE

The Complete Wine-Tasting Course
101 Wines You Need to Know

INTRODUCTION TO THE WINE-TASTING CLASS

Thirty years ago few wine schools catered to consumers, and the curricula of the schools that I attended were designed to teach facts and figures on viticulture, geography, grape varieties, regions, appellations, laws, wine making, and wine-making history. This was valuable information for aspiring professionals, but, as little time was devoted to teaching students how to taste wine, it was not of much use to beginners. Still, it was a compelling academic model and provided important information for anyone interested in gaining an in-depth understanding of wine, so I followed its basic curriculum when I wrote the original edition of *Windows on the World Complete Wine Course* in 1985. Over the past several years I have shifted priorities, believing that there is really only one way to learn about wine: pull the cork (or unscrew the screw cap!), pour the wine, smell, and taste. I now emphasize actual tasting, rather than learning facts, figures, maps, and vines.

I prepare a supplemental chapter with brand new material for each new edition of *Windows on the World Complete Wine Course*. Last year we realized that an important part of my class had never been addressed in this book. I'd never shared the list of wines we taste over eight classes with my readers, or why we taste them, the order in which we taste them, and how the wines help define the region from which they are produced. It's the most hands-on introduction to understanding the wonderful world of wine I know and it's included in this new edition, with completely updated wine lists, for your benefit. It's an important and valuable teaching tool and a fun way to taste and enjoy wine.

The wine lists follow the same tasting order I use for my eight-week wine course. For a true tasting education, I recommend you sample the wines in the same order. You'll begin with white wines, move on to red wines, and end with Champagnes, sparkling wines, and Ports. Each flight—or tasting—is carefully planned to emphasize specific characteristics of grape and region, from the simplest to the most complex. Work your way through each flight using the evaluation sheet and the 60-second wine expert tasting instructions on page 13 of this book.

I pour between eighty and ninety wines over eight classes at the Windows on the World Wine School, many of which are drawn from my personal cellar. I buy the rest from quality retail wine merchants whose wines are stored correctly and priced fairly. Buying these wines will cost you more than three thousand dollars, so I suggest you invite friends to participate in your own private wine school in order to share the expense and increase the fun. In my wine classes, each bottle is shared by sixteen students. Don't reveal the price of a wine until the wine or flight of wines has been tasted, evaluated, and voted on. It's more important to learn to taste with your senses than with dollar signs.

Because I work in New York City, the wine capital of the world, I have the luxury of being able to purchase any wine from any wine region in the world. You may not be able to find some of the exact wines on my class list so I've provided alternatives based on the style of the wine whenever possible. These options may be more readily available to you and should work just as well.

CLASS ONE: WHITE WINES OF FRANCE

Wines for Class One

1. Alsace: Trimbach Riesling 2005
2. Loire Valley: Muscadet, Marquis de Goulaine 2005 **BEST VALUE**
3. Loire Valley: Pouilly Fumé Jolivet 2005
4. Bordeaux, Graves, Pessac-Leognan: Château Carbonnieux 2005
5. Burgundy: Mâcon-Villages, Le Grand Cheneau 2005 **BEST VALUE**
6. Burgundy: Chablis Premier Cru, Vaillon, Moreau 2005
7. Burgundy: Meursault, Chanson 2005
8. Burgundy: Puligny-Montrachet, Les Folitières 2005, Olivier Leflaive
9. Burgundy: Corton Charlemagne, Bonneau du Martray 2002
10. Bordeaux: Sauternes: Château La Tour Blanche 1999

There are five parts to this tasting.

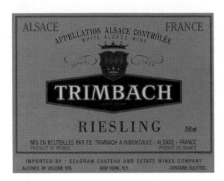

With this first wine we learn how to taste wine using the 60-second wine expert (p. 13). Pay attention to the balance of fruit and acid, and note that some high acid wines will taste better with food, especially shellfish.

PART I: ALSACE (ONE WINE, TASTED ALONE)

Wine #1: Trimbach Riesling 2005

This is the first wine of the course and I have selected it to set the tone for the next eight classes. It is a first-impression wine from the Alsace region, which has produced wines for hundreds of years. Riesling is their best grape. The Trimbach winery, established in 1626, is in its twelfth generation of family ownership and has a long history and tradition of producing high-quality wine.

I pour Trimbach Riesling first because it is unoaked and light in style, with the great fruit taste typical of this varietal. Alsace Rieslings generally have higher acidity, which turbo-charges everyone's palate.

Retail Price $19
Ready to Drink
Other recommended producers in Alsace are: Domaine Marcel Deiss, Domaine Weinbach, Domaine Zind-Humbrecht, Dopff "Au Moulin," Hugel & Fils, and Leon Beyer.

PART II: LOIRE VALLEY (TWO WINES, TASTED TOGETHER)

We now move our tasting into the Loire Valley with two totally different style wines, a Muscadet from the west coast of France, and a Pouilly Fumé from farther inland. Both are well-known French white wines. Neither wine is aged in wood so they are fruit friendly without any oak overtones.

Wine #2: Muscadet, Marquis de Goulaine 2005

The Muscadet is made with a grape variety called Melon de Bourgogne, a grape that originated in Burgundy, France, and transported to the Loire Valley. It should be tasted before the Pouilly Fumé because all Muscadets are lighter in style than Pouilly Fumés. Muscadet is a favorite wine in Parisian bistros because it is easy, light in style and has balanced fruit and acid. It is ready to drink when you buy it: the earlier the vintage, the better.

Retail Price $14 **BEST VALUE**
Ready to Drink
Other recommended producers for Muscadet are: Sauvion or Metaireau.

Wine #3: Pouilly Fumé, Jolivet 2005

Pouilly Fumé is made from 100 percent Sauvignon Blanc grapes. It is medium-bodied with more depth, complexity, and fruit concentration than the Muscadet. It also has a pungent aroma and bouquet that is instantly recognizable; some say that cat pee is an apt descriptor.

> Retail Price $24
> Ready to Drink
> Other recommended producers for Pouilly Fumé are: Michel Redde, Château de Tracy, Ladoucette, or Colin.

PART III: BORDEAUX (ONE WINE, TASTED ALONE)

Wine #4: Château Carbonnieux 2005, Bordeaux, Graves, Pessac-Leognan

Bordeaux now produces some of the world's best dry white wines, which has not always been the case. Bordeaux became famous for its high-end red wine and its extraordinary sweet wines (Sauternes). Over the last twenty years, the dry whites have achieved equal status to their superb reds and delicious sauternes.

The first three wines we tasted were aged and fermented stainless steel vats, made with a single grape variety, and each was produced in northern France. With Bordeaux, we move to the southwest of France, where the wines you'll taste are the result of blending, i.e., they are made with more than one grape variety. Château Carbonnieux is a blend of Sauvignon Blanc and Semillon and was aged in small French oak barrels.

Notice that the Carbonnieux has more color because:

1. It is older.
2. It has been aged in oak for at least six months.
3. It is produced in a southern geographic location.

All of those factors lead to a completely different taste experience. In the Carbonnieux, you'll find that the Semillon grape adds more weight (body), and the wine has less acid than the first three wines. The 2005 was a good vintage, not an exceptional one, but you'll experience the astringency of tannin,

If you can't find a Pouilly Fumé, use a Sancerre.

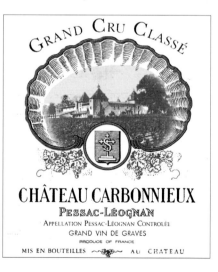

produced from its oak aging, for the first time. You should experience more tannin than fruit by the time you finish the 60-second wine expert, which is a sign that this wine will improve with age.

Retail Price $42
Ready to Drink, and can age
Other recommended producers for White Bordeaux are: Château Couhins-Lurton, Château La Tour-Martillac, Château Malartic-Lagravière, Château Smith-Haut-Lafitte, Château Bouscaut, Domaine de Chevalier, Château La Louvière, or Château Olivier.

PART IV: BURGUNDY (FIVE WINES TASTED TOGETHER)

In my opinion, Burgundy produces the best white wines in the world with a tremendous diversity of taste, weight, and style. The common denominator of the five wines in this flight is that they are all made from 100 percent Chardonnay grapes. You will sample unoaked, oak-aged, and barrel fermented Chardonnays in this tasting, from lightest to heaviest.

Pour the wines, then stand and look down at the five wines and note the different shades of color found in each. Those with more oak and age will have more color. Spend the next six to eight minutes concentrating on the aroma and bouquet of each wine. Decide which wine you think you will prefer based on color and smell alone. Once you've decided, discuss it with others or make notes on each. All professional wine tastings begin this way and in most international competitions many wines are never tasted but are rejected entirely because of their color and smell.

Even though I pour all Chardonnays together I break the tasting down into 2 sections:

Section I: Mâcon and Chablis: Unoaked
Section II: Meursault, Puligny-Montrachet, and Corton Charlemagne: Oaked

Section I: Unoaked wines

Wine #5: Mâcon-Villages, La Grand Cheneau 2005 (Mâconnais)

This is a medium-bodied, unoaked, balanced fruit acid wine in the best value category. It's always a crowd-pleaser.

> Retail Price $10 **BEST VALUE**
> Ready to Drink
> Other producers recommended are: Jadot, Duboeuf, and Bouchard.

Wine #6: Chablis Premier Cru, Vaillon, Christian Moreau 2005 (Chablis)

This wine is closer to the original three wines because it is also unoaked and, even though it is part of Burgundy, it comes from the northern limits, 115 miles from Paris. Cooler growing conditions and a limestone soil give us that tantalizing acidity that goes so well with shellfish and light styles of fish, such as sole and flounder.

> Retail Price $48
> Ready to Drink
> Other producers recommended are: Joseph Drouhin, La Schablisiene, William Fevre, and Robert Vocoret.

Section II: Oaked Wines

The next three wines are all from the Côte de Beaune region of Burgundy. The first two are from the most famous villages—Meursault and Puligny-Montrachet. The third is from a specific vineyard called Corton Charlemagne. They represent three different Burgundy quality levels. The Meursault is a village wine, the Puligny-Montrachet is a Premier Cru, and the Corton Charlemagne is a Grand Cru.

Wine #7: Meursault, Chanson 2005 (Côte de Beaune)

Depending upon the producer, a Meursault can range in style from medium to heavy. It is usually a fuller wine than Puligny-Montrachet and is one of the classic French wines usually included on French restaurant wine lists, and

often on American Contemporary wine lists. Chanson, which has a history going back to 1750, has gone through a metamorphosis over the last few years and is once again producing the style of wines I remember from early in my career. This wine has all the characteristics of a village wine: great fruit, acid, and oak balance, and it is accessible now.

Retail Price $60
Ready to Drink
Other producers recommended are: Bouchard Père & Fils, Joseph Drouhin, Labouré-Roi, Louis Jadot, Louis Latour, or Olivier Leflaive Frères.

Wine #8: Puligny-Montrachet, Olivier LeFlaive 2005, Les Folatières (Côte de Beaune)

One of my favorite wines, Puligny-Montrachet is a perfect compilation of what the French call *terroir*. This Puligny has tremendous fruit concentration with a subtle backbone of acidity because it's from a great producer and a great vintage. It's a wine that you can drink alone or one that will enhance nearly any meal. Les Folatières is one of the great vineyards located in the village of Puligny-Montrachet, and Olivier LeFlaive always produces a stylistically correct and high quality Puligny-Montrachet.

Retail Price $75
Ready to Drink, but can age
Other recommended producers are: Bouchard Père & Fils, Joseph Drouhin, Labouré-Roi, Louis Jadot, or Louis Latour.

Wine #9: Corton Charlemagne Bonneau du Martray 2002

Corton Charlemagne is one of the thirty-two Grand Cru vineyards and is also the largest of the Grand Cru white vineyards. It is barrel fermented as well as oak aged. This is an estate bottled wine versus a shipper's *négociant* wine, meaning all of the grapes are owned and the wine is produced at the Bonneau du Martray estate.

It is important when talking about ageability in wine tasting to show a wine that has gracefully matured with age. You will experience this instantly because the bouquet and aroma—toasty and elegant with a hint of soft fruits—are still going strong after six years.

ALL OF THE Bonneau de Martray wines are from vineyards rated Grand Cru. The vineyards are in the same location as they have been for the past 1,200 years.

Retail price $115
Ready to Drink
Other recommended producers for Burgundy wines are: Bouchard Père & Fils, Joseph Drouhin, Labouré-Roi, Louis Jadot, Louis Latour, Chevalier, Girardin, or Maison Champy.

PART V: BORDEAUX, SAUTERNES (ONE WINE, TASTED ALONE)

Wine # 10: Château LaTour Blanche 1999

Tasting this wine is truly a remarkable experience for anyone, and a great way to end the first class. The primary grape is Semillon. The viscosity, the smell of botrytis, the natural sweetness, and the long sweet acid finish leave everyone wanting more.

Retail Price $55
Ready to Drink
Other recommended producers for Sauternes are: Château Suduirat, Château Climens, Château Rieussec, Château Lafaurie-Peyraguey, or Château Doisy-Vedrines.

CLASS TWO: THE WHITE WINES OF CALIFORNIA

In the first class, we studied the white wines of France. In Class Two, we'll follow the same wine styles as with France, but this time we'll use American white wines.

Wines for Class Two:

1. Dr. Konstantin Frank Riesling Dry 2005 (Finger Lakes, New York)
2. Dry Creek Fumé Blanc 2005 (Sonoma County)
3. Mason Sauvignon Blanc 2005 (Napa Valley)
4. Hawk Crest Chardonnay 2006 (California) **BEST VALUE**
5. Blind Tasting
6. Blind Tasting
7. Chateau St. Jean Chardonnay 2005 (Sonoma County) **BEST VALUE**
8. Au Bon Climat Chardonnay 2005 (Santa Barbara)
9. Talbott Sleepy Hollow 2004 (Monterey)
10. Ramey Hudson Vineyards Chardonnay 2004 (Carneros)
11. Chateau Montelena 1999 (Napa)

There are six parts to this tasting.

PART I: RIESLING (ONE WINE, TASTED ALONE)

Wine #1: Dr. Konstantin Frank Dry Riesling 2005

In Class One, we started with a Riesling from Alsace. In this class we'll taste a Riesling from the Finger Lakes district of New York. Although this class is entitled White Wines of California, I believe the best American Rieslings are made in the Finger Lakes region of New York and in Washington State.

I want my first wine to excite the palate and get the saliva going, and this Dr. Frank Riesling is the perfect wine to do just that. It is an unoaked Riesling with a touch of residual sugar (1.4 percent) which balances perfectly with the crisp acidity of the Riesling fruit. The low alcohol (12 percent) does not mask any of the other components of the wine and leaves the palate refreshingly light.

Retail Price $20
Ready to Drink
Other recommended producers of this style of Riesling are: Hermann Weimer, Fox Run (both New York), and Chateau Ste. Michelle Eroica (Washington).

PART II: SAUVIGNON BLANC (TWO WINES, TASTED TOGETHER)

In Class One we tried two wines from the same region: the Loire Valley. In Class Two we are tasting two Sauvignon Blancs together, one from Sonoma and the other from Napa, to highlight the stylistic differences of this varietal.

Sauvignon Blanc is second only to Chardonnay in quality white wine grapes grown in California. Most Sauvignon Blancs are unoaked, light in style, have a perceived higher acidity, and for my tastes are more food-friendly than most Chardonnays. In fact, when I dine out, I usually choose a Sauvignon Blanc over a Chardonnay for my first course not only because they're lighter in style but also because they're far less expensive.

The following two wines are from two different AVAs. One is from Sonoma County and the other from Napa Valley; both are made with 100 percent Sauvignon Blanc. Both of these Sauvignon Blancs would be a perfect complement for any seafood appetizer.

Wine #2: Dry Creek Fumé Blanc 2005 (Sonoma County)

David Stare, the founder of Dry Creek Vineyards, was a pioneer in Sonoma, known especially for his work developing the Sauvignon Blanc grape. The AVA on this Dry Creek Sauvignon Blanc is Sonoma County, meaning that the grapes can come from anywhere within the county boundaries. The cooler climate of Sonoma County gives this wine lower alcohol content than our next wine. This wine has the distinctive Sauvignon Blanc bouquet with tangy grapefruit acidity.

Producers also have a choice when producing this varietal of either having the varietal name on the label or referring to it as Fumé Blanc. Dry Creek lists both on their label.

Retail Price $13.50
Ready to Drink
Other recommended producers of this style Sonoma County Sauvignon Blancs are: Simi, Ferrari-Carano, Mantanzas Creek, Chalk Hill, and Chateau St. Jean.

Wine #3: Mason Sauvignon Blanc 2005 (Napa Valley)

Because of its warmer climate, Napa Valley is God's gift to wine lovers—mostly for the world-class red wine made from Cabernet Sauvignon and Merlot. It's not necessarily the region I look to for Sauvignon Blanc, which generally thrives in cooler climates. With that said, Randy Mason makes one of the best Sauvignon Blancs in California. Students will immediately taste the great balance of the fruit and acid with mineral flavors.

> Retail Price $14
> Ready to Drink
> Other recommended producers of this style Napa Valley Sauvignon Blancs are: Robert Mondavi, Silverado, Phelps, and Caymus.

CHARDONNAY

So far we have tasted one Riesling and two Sauvignon Blancs. The next eight wines are made from the Chardonnay grape. Wine #4 is my French Mâconnais counterpart, an easy-drinking, light-style Chardonnay.

PART III: CHARDONNAY (ONE WINE, TASTED ALONE)

Wine #4: Hawk Crest Chardonnay 2006 (California)

Begin by looking at the label of this wine to find out where it comes from. The label states that the wine is from California, which means the grapes can come from any part of the state. The grapes for this Chardonnay are grown in the cool climate of Santa Lucia Highlands and are blended with grapes from Monterey County. This wine is styled after the Mâconnais region of France. Although the wine receives partial barrel fermentation, none of the oak overpowers the fruit. It is also aged *sur lee* and undergoes partial malolactic fermentation.

> Retail Price $11   **BEST VALUE**
> Ready to Drink
> Other recommended producers for this style Chardonnay are: Benziger, Estancia, Meridian, Rutherford Ranch, and Napa Ridge.

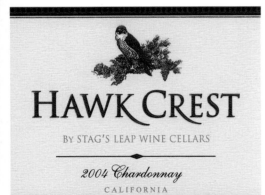

Although AVAs are important to understand, I still recommend that you should first buy wines based on the reputation of the producer. In this case, Stag's Leap Wine Cellars has a history and reputation for the production of great wine.

PART IV: CHARDONNAY
(TWO WINES, TASTED TOGETHER)

This is the first major blind tasting of the course (Wine #5 and Wine #6). I have always found in tasting wines that seeing a label is worth many years of experience. The only information I give the students is that both wines are made from 100 percent Chardonnay grapes. It's important to challenge yourself, as well as your guests, by testing your own abilities and thresholds of wine tasting. If you are conducting a tasting with friends, you should pour the wine first, without showing the label. Proceed with tasting by examining the color, smell, and taste of each wine, record your notes, and discuss the results with your friends.

Unfortunately, I can't tell you the exact wines I use since my students also get a copy of this book! Like any other blind tasting, this is a good exercise for anyone who wants to become better at wine judging. The wines should be served from lightest to heaviest.

The following are examples of some blind tastings that you can do at home:

Example 1: Unoaked California Chardonnay vs. Heavily Oaked
    California Chardonnay
Example 2: Sonoma Valley Chardonnay vs. Napa Valley Chardonnay
Example 3: White Burgundy vs. Napa or Sonoma Valley Chardonnay
Example 4: New York Chardonnay vs. California Chardonnay
Example 5: Australian Chardonnay vs. California Chardonnay

To be fair, both wines should be of the same quality, vintages within a year of each other, and in the same price range.

PART V: CHARDONNAY:
EXPLORING FOUR DIFFERENT AVAs
(FOUR WINES, TASTED TOGETHER)

The next four wines are Chardonnays produced in four different regions of California. Over the last thirty years, viticulturists have isolated the best soil and climatic conditions for growing the major grape varieties. For California Chardonnay, it is Sonoma, Monterey, Santa Barbara, and the Carneros region of Napa/Sonoma.

Just as in Class One, the four Chardonnays are poured out together. Once the wines are poured, stand up and look down at the colors of the wines. Before tasting anything, take the next five minutes to note the differences in color and smell among the four wines and decide which wines you think you will like the best.

Note on Wine #7 and Wine #8 that these wines should all be from the same vintage or within one year.

There are obvious differences in the style between the Chateau St. Jean and the Au Bon Climat Chardonnays. For years wine critics have said, "All California Chardonnays taste the same." By tasting these wines together, you'll understand that this criticism is not warranted.

Wine #7: Chateau St. Jean Chardonnay 2005 (Sonoma County)

The Chateau St. Jean winery is one of my favorite wineries of California. Since the early 1970s this producer has been making consistent top-quality wines at reasonable prices. In the early days it was mostly known for its production of white wines, and it continues to produce great Sauvignon Blancs and Chardonnays. It now produces Cinq Cepages, as well—one of the best red blends in California. This Chardonnay wine is barrel fermented with five months of *sur lie* aging and three quarters of the wine undergoes malolactic fermentation.

This wine is elegant, crowd-pleasing, and medium-bodied. It has perfect fruit oak balance and a lingering acidity. It is my number one pick for best quality value Chardonnay in California. The winemaker, Margo Van Staaveren, has been at Chateau St. Jean since 1979 and she knows how to make exceptional wines at all price levels!

Retail Price $14   **BEST VALUE**
Ready to Drink
Other recommended producers of this style "value" Chardonnay are: Gallo of Sonoma, Buena Vista, Sonoma-Cutrer (Russian River Ranches), and Kendall Jackson Vintner's Reserve.

Wine #8: Au Bon Climat Chardonnay 2005 (Santa Barbara)

In 2007, Au Bon Climat celebrated its twenty-fifth anniversary, making owner and winemaker Jim Clendenen one of the pioneers of Santa Barbara County winemakers. He has always produced a personal, idiosyncratic style of wine, similar to a French Burgundy, by choosing the best Chardonnay from specific

Simply put, malolactic fermentation means that a wine has undergone a two-step fermentation process. The secondary fermentation converts the stronger malic acid in wine into softer lactic acid. This adds complexity and lowers the acidity of wine. I always ask my students what the wine smells like. When it comes to Chardonnays, I often hear the word buttery (lactic). This is a sure sign that malolactic fermentation has taken place.

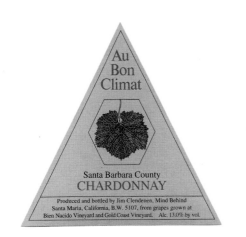

Au Bon Climat

Santa Barbara County
CHARDONNAY

Produced and bottled by Jim Clendenen, Mind Behind
Santa Maria, California, B.W. 5107, from grapes grown at
Bien Nacido Vineyard and Gold Coast Vineyard.   Alc. 13.0% by vol.

vineyards. This 2005 vintage Chardonnay comes from one of the top vineyards sites in California, the Bien Nacido Vineyard. This wine was fermented in small French Oak barrels, aged for another seven months *sur lie*, and bottled unfiltered after a year in barrel. The Au Bon Climat is a full-bodied, creamy, rich, and complex wine.

> Retail Price $16
> Ready to Drink, but will continue to age
> Other recommended producers of this Santa Barbara-style Chardonnay are: Sanford and Babcock.

The next two Chardonnays are both from the 2004 vintage and are from specific vineyards.

Wine #9: Talbott, Sleepy Hollow Vineyard Chardonnay 2004 (Monterey)

Talbott is one of the best wineries in Monterey County. They celebrated their twenty-fifth anniversary in 2007. Robert Talbott is one of the pioneers of high-quality winemaking in Monterey. The Sleepy Hollow Vineyard is located in the Santa Lucia Highlands AVA. All the grapes used in this Burgundian-style Chardonnay come from the oldest part of the Sleepy Hollow Vineyard. This is a ripe, rich, complex wine that has been 100 percent barrel fermented and barrel aged for one year *sur lie*. The toastiness of the oak blends in perfectly with the ripeness of the grapes. The biggest difference between the Chateau St. Jean, Au Bon Climat, and this Talbott Chardonnay is the alcohol content. It jumps to 14.7 percent, more than one degree higher than the others, creating much more density and weight in the wine.

> Retail Price $45
> Continues to age
> Other recommended producers of this Monterey-style Chardonnay are: Mer Soleil, Chalone, Morgan, and Calera.

Wine #10: Ramey, Hudson Vineyard Chardonnay 2004 (Carneros)

David Ramey has been one of the top winemakers in California for many years. I enjoyed tasting his earlier wines when he was a winemaker at the Chalk Hill and Dominus wineries. Two of the vineyards from Carneros that David has worked closely with over the years are the Hyde Vineyard and the

Talking to winemakers on the subject of "to filter or not to filter" always leads to great discussions and remains a controversial subject today. Filtration is the process by which solids in the wine are removed. By filtering a wine, you take out any impurities that can make the wine unstable. Many winemakers feel that filtering a wine will take away its varietal character and complexity. An unfiltered wine, especially after age, will throw sediment both in red or white wine.

Sur lie, a French term, is used when a wine is left aging with its sediment, such as dead yeast cells and grape skins and seeds.

Hudson Vineyard. They are both from the Carneros region, which has one of the coolest climatic conditions of the North Coast Counties, perfect for the Chardonnay grape.

This wine from the Hudson Vineyard produces a powerful full-bodied Chardonnay. The 2004 Hudson Vineyard grapes were whole-clustered pressed and fermented in small French Burgundian barrels (70 percent new barrels) using native yeast. At 14.8 percent alcohol content, this is a big style, in-your-face Chardonnay that will benefit from many more years of aging. Like the previous three Chardonnays, this wine was also aged *sur lie*, but for twenty-one months and bottled without filtration.

Retail Price $48
Will benefit from further years of aging
Other recommended producers of this Carneros-style Chardonnay are: Kistler Vineyards Hudson Vineyard, Kistler Vineyards Hyde Vineyard, and Williams Selyem Allen Vineyard.

PART VI: TASTING AN AGED CALIFORNIA CHARDONNAY (ONE WINE, TASTED ALONE)

One of the most common questions raised in my class is the ageability of wine. Our last Chardonnay is more than eight years old; it is extremely important that students understand what happens to a California Chardonnay as it gets older. This next wine well demonstrates the effects of age on wine.

Wine #11: Chateau Montelena 1999 (Napa Valley)

This winery was originally established in 1882 and was resurrected by the Barrett family in 1972, about the time I seriously began my wine career. We've grown up together, which is why it's always been one of my sentimental favorites. Chateau Montelena became one of the most famous wineries in California in 1976 when their 1973 Chardonnay took first place in a comparative tasting of California Chardonnays and French White Burgundies. This upset was so shocking, so revolutionary, that a book, *Judgment in Paris*, was written about the tasting, and it was recently announced that a movie is being planned. The winemaker at that time was Mike Grgich, who now produces exceptional wines at his own winery, Grgich Hills.

This is a textbook-style Chardonnay conforming to every descriptive word critics write about great Chardonnays: great balanced fruit, and a long, complex aftertaste with refreshing acidity. After eight years of age, the tannins have mellowed, the wine is the softer, fruit becomes the dominant factor, yet there is still enough acidity in this wine for it to age another five years.

The big difference between this Chardonnay versus the other Chardonnays you have just tasted is that the Chateau Montelena does not undergo malolactic fermentation. This is a village-style Burgundian wine, elegant with perfectly integrated acid/fruit levels. It is also aged on its yeast and has spent eight months in oak.

Retail Price $115
Ready to Drink, but will continue to age gracefully
Try finding a Chardonnay with some age, at least five years, in order to see how aging affects this type of wine.

CLASS THREE: WHITE WINES OF GERMANY

The Wines of Germany is the third class and the final white wine tasting before we move on to the reds. It is divided into two parts, a component tasting and then the white wines of Germany.

In the component tasting, you will learn how to identify the four critical components of wine—acid, sugar, tannin, and sulfites—by adding each to water, then to wine. You will get to know your own tolerance thresholds for each as you experience the individual components. This is the ideal introduction to German wines. Why? Because I describe German wine as the perfect balance of residual sugar, fruit, and acid.

White Wines of Germany highlights the stylistic differences between the Mosel and Rhein regions. The wines of the Mosel are generally lower in alcohol and lighter in style than those of the Rhine region.

The real point, of course, is exposing you to the basic characteristics of German white wines.

Part I: Prelude: Component Tasting

What you'll need:

1. Control wine: I recommend either a Gallo Chablis Blanc or a Carlo Rossi Chablis
2. 9 wineglasses
3. Component chart
4. Component tasting worksheet
5. Acids found in wine (a blend of all three, or just one): tartaric, citric, and malic
6. Sugar (superfine)
7. Tannin
8. Sulfur dioxide (sold as potassium metabisulfite in home winemaking shops)

This component tasting will help you understand your own olfactory thresholds for taste (residual sugar and acid), tactile sensation (tannin), and smell (fruit and sulfur dioxide). If you want to try this component tasting at home you can purchase all of the components at home winemaking stores. You'll need nine glasses: one for the control wine and eight for the components. Whichever Control Wine you select, use that wine for all of the component tastings.

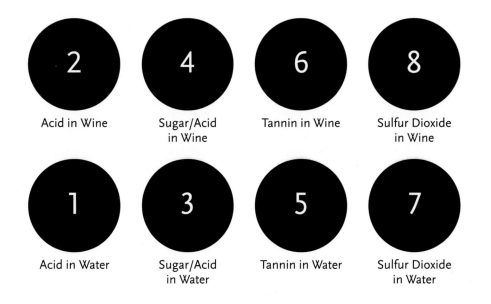

| 2 | 4 | 6 | 8 |
| Acid in Wine | Sugar/Acid in Wine | Tannin in Wine | Sulfur Dioxide in Wine |

| 1 | 3 | 5 | 7 |
| Acid in Water | Sugar/Acid in Water | Tannin in Water | Sulfur Dioxide in Water |

COMPONENT TASTING WORKSHEET

Control Wine:

60-Second Wine Expert:

Identify the major component in each time slot.

0–15 seconds _____   30–45 seconds _____

15–30 seconds _____   45–60 seconds _____

Component I:

1. Acid in water: _____

2. Acid in wine: _____

Component II:

3. Sugar/acid in water: _____

4. Sugar/acid in wine: _____

Component III:

5. Tannin in water: _____

6. Tannin in wine: _____

Component IV:

7. Sulphur dioxide in water: _____

8. Sulphur dioxide in wine: _____

Control wine

Taste the control wine and record the balance of the wine's components. Did the wine have any residual sugar? Was the wine low, medium, or high in acidity? Was the wine light-, medium-, or full-bodied? Were its components all in balance or was there one that dominated? Although everyone starts with the same wine, it's amazing to see how each taster's threshold for each component differs. In last semester's class one third of the students rated the acidity as high, one third felt it was balanced, and another third judged it as low, all of which highlight the fact that everyone has different "taste" thresholds!

Once you've finished discussing the control wine, continue with the component tasting, which is divided into four sections, one for each component.

Component I: Adding acid

Acid in Water: Add 1½ teaspoons of acid mixture (tartaric, citric, and malic acids) to 1 liter of water.

Pour two ounces of the acid/water blend for each taster and taste; record and compare your reactions. Acidity is usually most strongly felt on the sides of the tongue and in your gums. Pay particular attention to these areas of your mouth as you taste the first two glasses in this trial. By isolating the components in neutral water, you will experience acidity, nothing else, which instantly highlights your acidity threshold: Some people cringe at the sourness while others find it palatable, even pleasant.

Now comes one of the most important lessons of the tasting class: the order in which the wines are served and how one wine can affect another. Go back to the control wine and taste it again. Has the taste changed? Most likely your answer will be an emphatic *yes*! However, the wine hasn't changed, your awareness of acidity has. The high acid water component was so strong that it affected the taste of the control wine, which now seems more acidic than it did on your first taste.

Acid in Wine: Add 1 teaspoon of acid mixture to 1 liter of wine.

Pour a two ounce serving of the Acid/Wine mix for each taster. By raising the acid level of the wine you've obviously changed the balance of the wine since acidity is wine's strongest component. Taste the wine and record your notes. In class, I've found some students prefer this style over the control wine, while others wanted their money back! Additionally, this tasting will help you understand how adding acid to wine is one way winemakers can affect a wine's style.

COMPONENT II: ADDING SUGAR AND ACID

Sugar and acid to water: Add 8 teaspoons of superfine sugar and l teaspoon of acid to one liter of water.
Sugar and acid to wine: Add 6 teaspoons of superfine sugar and l teaspoon of acid mixture to l liter of wine.

Sweetness is usually tasted on the tip of the tongue; experiment with how your tongue reacts to sweetness. After trying the sugar/acid mixture in water, try the sugar/acid wine. Record your thoughts of both and discuss, then go back and taste the control wine. Since your tongue has adjusted itself to the double whammy of both higher acidity and higher sugar, you'll find the balance of the components in the control wine has completely changed.

COMPONENT III: ADDING TANNIN

Tannin is sold in winemaking stores in a powder form. In order to make the tannin solution, mix 3 teaspoons of tannin powder with 6 ounces of warm or hot water, as tannin will not dissolve in cold water.

Tannin to water: Mix l teaspoon of tannin solution into l liter of water.
Tannin to wine: Add l teaspoon of tannin solution to l liter of wine.

Tannin affects the balance and taste of wine. As before, taste and discuss the effects to adding tannin to water, then to the control wine. Take notes on each. You'll discover that the astringency of the tannin in both the water and the wine will overpower and dry out your taste buds; your mouth may become so dry that you will lose your saliva, a key element in taste. By this time in the component tasting, the control wine has undergone many "changes" and few traces of its original taste remain. The tannin on your tongue will overpower the sweetness and fruit of the control wine, leaving only acidity and tannin.

Component IV: Adding sulfur dioxide (SO₂)

I'm frequently asked questions about the use of sulfites in wine, usually from people asking why sulfur dioxide (potassium metabisulfite) is added to a wine and if it is healthy. First, all wines contain some SO_2 because it is a natural by-product of fermentation. Second, because SO_2 kills bacteria in wine, prevents unwanted fermentation, and acts as a preservative, it is occasionally added during the winemaking process (see page 10). Many people confuse sulfur dioxide with hydrogen sulfite, which has the smell of rotten eggs; sulfur dioxide is more of a sensation than a smell. Too much sulfur dioxide in wine can produce an itching sensation in your nose, sneezing, and watering of the eyes. Is it healthy? Not really, but it's not unhealthy, either, unless you have a natural allergy to it. Because it is a naturally occurring component in wine, it's important that you understand what your tolerance for it is. I have an extremely low threshold for sulfur dioxide and those who are allergic may have a stronger reaction.

Sulfur dioxide to water: Add ⅛ teaspoon of SO_2 to l liter of water.
Sulfur dioxide to wine: Add ⅛ teaspoon of SO_2 to l liter of wine.

Add the same amount of SO_2 to the water as you do to the wine; first smell the water, then the wine. It is not neccessary to taste it. Interestingly most students do not experience the sulfur dioxide in water, but, due to the alcohol content, many of those same students will understand SO_2 once it's added to the control wine and inhaled; still some won't experience it at all. Individual tolerance depends on personal thresholds. You may find it repugnant (as I do); it may not bother you at all; or you might fall somewhere in between. In any event, if sulfur dioxide dominates a wine, the wine is flawed.

Part II: White Wines of Germany

At this point in my class, students need some *real* wine. The refreshing fruit, sweetness, and acidity of good German white wines provide the perfect transition.

I've found that German wines are not as well-known or as well-understood as French or California wines to most Americans. German wines are often dismissed because they seem complicated, and also because they are light-bodied

In my class, I pre-pour the first 6 of the 8 water/wine components. When my students enter the classroom they often notice that the room has a pleasant wine cellar smell. If I had pre-poured the sulfur dioxide, the classroom would smell more like a laboratory.

In Germany, 100,000 grape growers cultivate nearly 270,000 acres of vines, meaning the average holding per grower is 2.7 acres.

white wines and the best German wines have higher levels of residual sugar. Don't make that mistake. Once you learn the main villages, vineyards, classifications, and a little German pronunciation, you'll find it easy to understand and you'll be able to enjoy the charm and elegance of the great German wines.

Here are the wines I poured in my most recent class:

1. A blind wine: Any Qualitätswein from Germany
2. Graacher Himmelreich Riesling Kabinett 2005, J.J. Prum (Mosel)
3. Niersteiner Bruckchen Riesling Kabinett 2005, Strub (Rheinhessen)
4. Wehlener Sonnenuhr Riesling Spatlese 2005, Meulenhof (Mosel)
5. Gleisweiler Holle Riesling Spatlese 2005, Minges (Rhein-Pfalz)
6. Urziger Wurzgarten Riesling Auslese 2003, J. J. Christofffel Erben (Mosel)
7. Deidesheimer Grainhubel Riesling Auslese 2001, Dr. Deinhard (Rhien-Pfalz)
8. Lieser Niederberg-Helden Riesling Beerenauslese 1999, Sybille Kuntz (Mosel)

There are five sections to this tasting.

SECTION I: IDENTIFYING THE CHARACTERISTICS OF A GOOD QUALITY GERMAN WINE (ONE WINE, TASTED ALONE)

Wine #1 should be tasted blind if you have several people participating in your tasting. I chose a good typical Qualitätswein because it is medium-dry and has a good balance of fruit and acid with a touch of residual sugar. Qualitätswein typifies German wine, so taste it as your control wine, take notes, and pay attention to the fruit, acid and sugar. As noted on page 87, there are three different levels of German wines moving from the basic classification Tafelwein, or table wine, to Qualitätswein, to the highest classification Präedikatswein.

Retail price $15
Ready to Drink

The next seven wines are ordered accordingly. I believe the highest quality German wines are made with the Riesling grape variety and all of the Präedikatswein recommended for this tasting are produced from 100 percent Riesling.

Five great vintages in a row, 2001 to 2005, were exceptional vintages in both the Mosel and the Rhein regions of Germany, a very rare occurrence because of their northerly locations.

Section II: Tasting and Comparing Two Kabinetts
(two wines, tasted together, wines should be the same
vintage or no more than one year apart)

Wine #2: Graacher Himmelreich Riesling Kabinett 2005, J.J. Prüm (Mosel)

Wine #2 comes from the Mosel region, the village of Graach and the Himmelreich vineyard. It is classified a Kabinett based on its residual sugar content. The Prüm family owns some of the best vineyard land in all of the Mosel. It is made from 100 percent Riesling and comes from the outstanding 2005 vintage. This Riesling will be the driest and lightest wine of the next six wines and usually has a little *spritzig*, a slight bit of carbon dioxide purposely leftover after fermentation, enhancing the flavor of the refreshing and lively Riesling fruit.

> Retail Price $25
> Ready to Drink
> Other recommended producers of this style Mosel Kabinett are: Kesselstatt, Bergweiller-Prum, Jos Christoffel, and Frederich-Wilhelm-Gymnasium.

Wine #3: Niersteiner Brückchen Riesling Kabinett 2005, Strub (Rheinhessen)

The Niersteiner Brückchen Riesling Kabinett 2005, Strub comes from the Rheinhessen region, the village of Nierstein and the Brückchen vineyard and is classified a Kabinett. Walter Strub is the 11th generation winemaker. All the grapes for this wine are hand-picked, whole bunch pressed, and fermented slowly. This wine has more concentration, body, and depth of Riesling flavor than the Graacher, and a vibrant, balanced structure.

> Retail Price $17
> Ready to Drink
> Other recommended wines of this style Rheinhessen Kabinett are: Niersteiner Paterberg, Niersteiner Oelber, or Niersteiner Orbel, all from the producer Strub.

Section III: Tasting and Comparing Two Spätlese
(two wines, tasted together, wines should be of the same vintage or within one year)

Wine #4: *Wehlener Sonnenuhr Riesling Spätlese 2005, Meulenhof (Mosel)*

This wine is from the Mosel region, the village of Wehlen, and from one of the most famous vineyards in Germany-Sonnenuhr, and is classified a Spätlese. The estate has a history going back into the twelfth century. Even though this wine has higher residual sugar content than the Kabinetts, the delicate Riesling fruit and zesty acidity gives this wine a lingering finish.

> Retail Price $30
> Ready to Drink, but can age
> Other recommended producers of this style of Mosel Riesling Spätlese are: Selbach-Oster, J.J. Prum, Kesselstatt, S.A. Prum, Kerpen.

Wine #5: *Gleisweiler Holle Riesling Spätlese 2005, Minges (Rhein-Pfalz)*

This wine is from the Rhein-Pfalz region, the village of Gleisweil, and from the Holle vineyard and is classified a Spätlese. The history of this winery is traced back to the fifteenth century and the Minges family ownership is in its sixth generation. The Minges family also owns a fairly large vineyard holding in the Pfalz region. The exotic bouquet combined with its chalkiness and mineral overtones give this Riesling a well-rounded finish.

> Retail Price $27
> Ready to Drink, but can age
> Another recommended producer of this style of Rhein-Pfalz Riesling Spätlese is: Weingut Johannishof.

The Rheinhessen, Rheingau, and the Rhein Pfalz all produce different style Rieslings coming from different soils, slopes, and proximity to the rivers.

Section IV: Tasting of Two Auslese
(two wines, tasted together, not necessary to have the same vintage, but must show younger to older)

Wine #6: *Urziger Wurzgarten Riesling Auslese 2003, J.J. Christoffel Erben (Mosel)*

This wine is from the Mosel region, the village of Urgiz and the Wurzgarten vineyard, and is classified an Auslese from the exceptional 2003 vintage. Once we get to the German Auslese classification it's a whole new ballgame. We are now into what I would describe as a dessert wine category. The *weingut* of J.J. Christoffel Erben is small, only producing 2,500 cases a year, all very high quality. This is a wine of great finesse. The immediate sweetness followed by the tropical fruit and juicy acidity give pleasure from first to last taste.

Retail Price $29
Ready to Drink, but can age
Other recommended producers of this style of Mosel Riesling Auslese
are: Dr. Loosen, Karl Erbes, and Alfred Merkelbach.

Wine #7: Deidesheimer Grainhubel Riesling Auslese 2001, Dr. Deinhard (Rhien-Pfalz)

Weingut is the German name for winery.

This wine is from the Pfalz region, the village of Deidesheim, the Grainhubel vineyard, and is classified an Auslese. Deinhard was founded in 1849 and today is one of the largest producers of Sekt (sparkling wine) in Germany. With over 35 hectares of vineyards, it is the largest estate of all tasted in this class. This Auslese has a drier style than the Urziger, but has a nice harmony of components, dominated by the intensity of the Riesling fruit. The extra two years of aging demonstrate that many German wines have the potential to age.

Retail Price $45
Ready to Drink, but can age
Other recommended producers of this style of Rhein-Pfalz Riesling
Auslese are: Bassermann-Jordan and Burklin-Wolf.

SECTION V: ONE OF THE GREATEST SWEET WINES OF THE WORLD: BEERENAUSLESE (ONE WINE, TASTED ALONE)

Wine #8: Lieser Niederberg-Helden Riesling Beerenauslese 1999, Sybille Kuntz (Mosel)

This wine is from the Mosel region, the village of Lieser and the Niederberg-Helden vineyard, and is classified a Beerenauslese. This final wine, a Sybille Beerenauslese, typifies the love/hate relationship many Americans have with German wines. It's high sugar content (9 percent) is both seductive and delicious. It's sweet, but intoxicating because of its near perfect balance of fruit, acid, and residual sugar. This is a lighter style Beerenauslese which is beginning to show some maturity. The botrytis brings out the richness of honeyed apricot flavors.

Retail Price $150
Ready to Drink, but can age
Other recommended producers of this style of Mosel Beerenauslese are: Fritz Haas, Dr. Loosen, Egon Muller, J.J. Prum, and Schloos Lieser.

Finding a good Beerenauslese is getting harder and harder, even in the great wine stores in New York City. Over the past few years, I have substituted for a Beerenauslese a Hungarian Tokaji Aszu, 5 Puttonyos. The Tokaji is also one of the great sweet wines of the world.

CLASS FOUR: RED WINES OF BURGUNDY AND THE RHÔNE VALLEY

One of the most difficult subjects to teach is the wines of Burgundy. At the beginning stages of the *Windows on the World* book, my publisher wanted to be convinced that I could write a "simple" guide to wine. He asked me to choose the most difficult wine region to understand and write a simple chapter. I chose Burgundy. He liked what I wrote and the book became a reality. But writing that chapter wasn't easy.

The three most important things to know about Burgundy wine, ranked in order of importance, are: the producer, the vintage, and the classification system. Each of these differences becomes more evident after tasting each flight.

Some of my favorite wines come from the Rhône Valley. From the simple, medium-bodied Côte du Rhône, to the spicy Crozes Hermitage and Hermitage, to the big, voluptuous, high-alcohol Châteauneuf-du-Papes, these are some of the greatest wines in the world and still represent one of the best values in all winedom. If I ever add another wine class to the school and to the book, it would be devoted to wines made from the Grenache and Syrah grape varieties, and leading the way would be the wines of the Rhône Valley.

Wines for Class Four:

1. Beaujolais Villages, Louis Jadot 2005
2. Fleurie, Georges Duboeuf 2005
3. Mercurey, Domaine de la Croix Jacquelet, Faiveley 2003

4. Pommard, Labouré-Roi 2002

5. Volnay-Santenots, Domaine Ballot-Millot et Fils 2002

6. Chambolle-Musigny, Joseph Drouhin 1999

7. Nuits St. Georges, Les Haut Pruliers, Domaine Daniel Rion 1999

8. Clos Vougeot, Louis Latour 1999

9. Crozes Hermitage, Les Jalets, Jaboulet 2003 **BEST VALUE**

10. Châteauneuf du Pape, La Bernardine, Chapoutier 2003

There are five parts to this tasting.

PART I: THE WINES OF BEAUJOLAIS (TWO WINES TASTED TOGETHER, AND OF THE SAME VINTAGE)

When it comes to Burgundy, I start off with the lightest of the red wines: those of the Beaujolais region and made from 100 percent Gamay grapes. These two wines will enhance your understanding of just how important the Burgundy wine producers, or négociants, are. I begin with Beaujolais Villages since it is a lesser appellation than the Fleurie.

Pour the two wines; stand and study both from above. Do you notice a difference in color? Does one wine have a deeper color than the other?

You should notice a distinct difference in color: the Beaujolais Villages is lighter than the Fleurie even though both wines were made using 100 percent Gamay grapes and both were produced in 2005, a great vintage. This difference in color tells us a lot about the two wines: lighter colors indicate higher acidity, lower tannins, and demonstrates its readiness or drinkability.

Color is dictated by a wine's classification and the winemaker's technique. Because the color of wine derives entirely from the skin of the grape, the longer a wine ferments with its skin—in a process called "maceration"—the darker its color will be. In our case, the darker red of the Fleurie tells us that it comes from a higher concentration of grapes that spent more time macerating with its skins than did the Villages.

Wine #1: Beaujolais Villages, Louis Jadot 2005

The Beaujolais Villages Jadot smells like fresh grape juice. It reminds me of the smell of a winery during harvest. It has lots of red fruit with raspberry and strawberry smells, very little tannin, and high acidity. The grapes for this wine can only come from designated villages within the Beaujolais district.

Retail Price $10
Ready to Drink
Other recommended producers of the Beaujolais Village style wine are: Bouchard, Drouhin, and DuBoeuf.

Wine #2: Fleurie, Georges Duboeuf 2005

The Beaujolais Fleurie DuBoeuf 2005 has a more concentrated Gamay bouquet, more tannin, higher alcohol, more fruit, and less acidity than the Beaujolais Villages. The grapes from this wine can only come from the village of Fleurie.

Retail Price $13
Ready to Drink
If you can't find the Beaujolais from the village of Fleurie try another village such as Brouilly or Moulin-À-Vent. Other recommended producers are: Joseph Drouhin and Louis Jadot.

I use the producers Louis Jadot and Georges DuBoeuf because they are both high-quality wines and also the most widely distributed Beaujolais in the United States.

PART II: THE WINE OF THE CÔTE CHÂLONNAISE (ONE WINE, TASTED ALONE)

Wine #3: Mercurey, Domaine de la Croix Jacquelet, Faiveley 2003

We move now to Pinot Noir, whose bouquet and aroma bear little resemblance to that of the Gamay. Pinot Noir is more concentrated, spicy, and seductive. The majority of the Burgundy class is devoted to the wines of the Côte d'Or but I would be remiss were I not to include a tasting of a wine from the Côte Châlonnaise. I choose this Pinot Noir from Burgundy because it is easy to drink, accessible now, will be the lightest and least expensive of all of the Pinot Noirs that we will taste. It also provides an excellent transition between the wines of Beaujolais and the Côte de Beaune, which follows.

The 2003 vintage in France is known for the extreme heat of the year, which increases the sugar content of grapes, shortens the growing season, and makes it more difficult for growers to predict perfect harvest times. This wine is made from grapes picked at their perfect ripeness. It is more

complex, fuller in body, and has more integrated fruits and tannins than any of the Beaujolais wines.

> Retail Price $18
> Ready to Drink
> Other recommended producers of the Mercurey style wines are: Domaine de Suremain and Michel Juillot.

The next five wines are all from the Côte d'Or and I begin with two wines from the Côte de Beaune and then continue with the three wines from the Côte de Nuits. Wines from the Côte de Beaune have historically been lighter in style than those from the Côte de Nuits.

PART III: A VILLAGE WINE AND A PREMIER CRU FROM THE CÔTE DE BEAUNE (TWO WINES, TASTED TOGETHER, PREFERABLY OF THE SAME VINTAGE)

The Pommard is a village wine, while the Volnay is from a Premier Cru vineyard: The Pommard is tasted first. Tasting these two wines together highlights the differences between a village and a Premier Cru vineyard. The year 2002 was a great vintage in Burgundy.

Wine #4: Pommard, Labouré-Roi 2002

The grapes for this wine can only come from village of Pommard. This Pommard is a well-made, classic-style Pinot Noir with a medium amount of fruit, tannin, and a great balance. It is the perfect restaurant Burgundy, accessible now and reasonably priced.

> Retail Price $30
> Ready to Drink
> Other recommended négociants for Côte de Beaune style wines are: Bouchard Père & Fils, Joseph Drouhin, Jaffelin, Louis Jadot, and Louis Latour.

What is the difference between a négociant and an estate bottled wine? A négociant is a cooperative and buys the grapes or wine directly from a specific appellation. The négociant then bottles the wine they've either bought or made under the négociant name, such as Labouré-Roi. A domaine or estate bottled wine means that the grapes from the village or vineyard listed on the label are owned and grown, and the wine is produced at that property, such as with Ballot-Millot. Most négociants offer a full selection of wine at all different price and quality levels throughout Burgundy.

Wine #5: Volnay-Santenots, Domaine Ballot-Millot et Fils 2002

The grapes for this wine can only come from the vineyard of Santenots located in the village of Volnay. Domaine Ballot-Millot specializes in high-quality Premier Cru vineyards in the Côte de Beaune. This is an artisanal-style wine which not only shows the charm of the Pinot Noir grape but of the *terroir*, a French word that loosely translates to mean the combination of soil, climate, and slope of the vineyard site. When I was twenty years old I tasted my first Volnay: I shall never forget its finesse, elegance, and subtlety.

> Retail Price $50
> Ready to Drink
> Other recommended vineyards in Volnay are: Callieret, Champans, and Clos des Chénes. Other recommended Domaine or estate bottled producers of the Volnay style wines are: Domaine Clerget, Drouhin, Comte Lafon, and Marquis d'Angerville.

PART IV: THE CÔTE DE NUITS: A VILLAGE, A PREMIER CRU, AND GRAND CRU (ALL THREE TASTED TOGETHER; IT IS MOST IMPORTANT THAT WINES #6 AND #7 ARE THE SAME VINTAGE, WHILE WINE #8 CAN BE THE SAME VINTAGE OR OLDER)

Thus far, I've discussed the importance of producers, vintage year, and the different classifications of Burgundy wine (Village wine, Premier Cru, and Grand Cru). The only real way to understand the differences between these classifications and the Burgundy appellations, however, is to taste them together. The next three wines represent some of the best reds produced in Burgundy.

The Côte de Nuit is God's gift to Pinot Noir lovers, and these great Burgundies show the essence of the Pinot Noir taste. A recent class was extremely fortunate in having three wines from the great 1999 vintage selected from my personal wine cellar. All three wines are from three outstanding producers: Louis Latour, Joseph Drouhin, and Daniel Rion. You will taste the differences among the village wine of Chambolle-Musigny, a Premier Cru from Nuits St. George, and the Grand Cru Clos Vougeot. All three are exceptional in their own right, but you should pay particular attention to the differences in appellations and style. As you taste your way through the wines the differences will become very apparent.

I'm going to begin this tasting by asking you to pour the next five wines— the three Burgundy wines (#6, #7, and #8) in this flight, and the two Rhône Valley wines (#9 and #10). Examine the colors of each. Notice the difference between the lighter colors of the Pinot Noir-based Burgundies and the deeper colors of the Grenache/Syrah-based Rhône Valley wines. The difference in color provides a quick visual demonstration of the differences in texture, intensity, body. and color of the two regions' wines.

Set the two Rhône Valley wines aside and proceed with the three Burgundy wines in this flight.

Wine #6: *Chambolle-Musigny, Joseph Drouhin 1999*

The Pinot Noir grapes can only come from the village of Chambolle-Musigny. The wine firm of Joseph Drouhin, established in 1756, produces both négociant wines and domaine wines, and is considered one of the top names in all of Burgundy by wine collectors and professional buyers. I begin with this wine because of its Village wine appellation, which makes it the lightest in style of the three wines. This medium-bodied Chambolle-Musigny is dominated by red fruits and spices with all of the elements of a great Pinot Noir: intensity of fruit, elegance, and a delicate balance of all of its components. It is a superb example of what a village wine from Côte de Nuits is all about.

> Retail Price $44
> Ready to Drink
> Other recommended producers of the Chambolle-Musigny style wines are: Louis Jadot, Georges Roumier, and Robert Groffier.

Wine #7: *Nuit St. Georges Les Haut Pruliers, Domaine Daniel Rion 1999*

All of the Pinot Noir grapes for this wine come from the Les Haut Pruliers vineyard in the village of Nuit St. Georges. Daniel Rion is one of the top Domaines of Nuits St. Georges.

The wines of Nuits St. Georges have a different underlying structure than other villages of the Côtes de Nuits. This wine is dark in color, full, powerful, firm, concentrated, and tannic—very different in style to the delicate Chambolle-Musigny. This is the Pinot Noir for those who prefer Cabernet Sauvi-

MIS EN BOUTEILLE   AU DOMAINE

NUITS-ST-GEORGES
LES HAUTS PRULIERS
PREMIER CRU
1999
DOMAINE DANIEL RION & FILS
21 PREMEAUX ◆ FRANCE

gnon. In tasting Wines #6 and #7 together, you'll experience the diversity and range of the Pinot Noir grape wines produced within the Côte de Nuits region.

Retail Price $45
Needs more time
Other recommended producers of the Nuit St. Georges style wines are: Faiveley, Henri Gouges, Jean Grivot, and Henri Jayer.

Wine #8: Clos Vougeot, Louis Latour 1999

All the Pinot Noir grapes for this wine come from the Grand Cru vineyard Clos de Vougeot. This is the largest Grand Cru vineyard in Burgundy with nearly 124 acres and 77 owners/producers. This Clos de Vougeot is made by the firm Louis Latour, who makes the viticultural decisions on how to grow the grapes, when to pick them, what fermentation process the wine shall undergo, and how long the wine will age before its release. The other 76 owners make those decisions for their wines.

Of all of the Grand Crus of the Côte de Nuits, Clos de Vougeot is the lightest in style, lowest in tannin, and still accessible when young. The subtlety, elegance, and charm of this wine with black fruits such as currants and cherries seduce tasters into believing it to be ready to drink, yet it will improve for as long as ten more years if it is properly aged. Because of the amount of wine produced, this Clos de Vougeot is an affordable way to try a Grand Cru Burgundy. If you like this wine, buy yourself a case, store it properly, and dream on the joy of drinking it in ten years' time.

Retail Price $97
Ready to Drink
Other recommended producers of the Clos de Vougeot style wines are: Leroy, Méo-Camuzet, Mongeard-Mugneret, George Roumier, and Jean Grivot.

PART V: THE CÔTE DU RHÔNE (TWO WINES, TASTED TOGETHER)

After going through all of the major villages, Premier Crus, and Grand Crus of Burgundy, my students and I are thoroughly exhausted and I usually have a mere twenty minutes to cover one of my favorite regions on earth—the

Rhône Valley. Fortunately, the Rhône Valley is far easier to understand than Burgundy.

Compare the lighter colors of Pinot Noir-based Burgundies to the deeper colors of the Grenache/Syrah-based Rhône Valley wines. You'll get a quick picture of the differences in texture, intensity, and body, as well as color.

I begin with the lesser appellation Crozes-Hermitage and end with the higher appellation of Châteauneuf du Pape. Both wines are from the same 2003 vintage, which was also the year of a really hot summer in France. The hot sun, because it forced the grapes to produce more sugar, gave this vintage higher alcohol levels than normal.

Wine #9: Crozes-Hermitage, Les Jalets, Jaboulet 2003

This wine comes from the northern part of the Rhône Valley and is made with 100 percent Syrah. Jaboulet is one of the Rhône Valley's greatest producers. I personally consume a lot of this wine, finding that it is neither as light as a Pinot Noir nor as heavy as a Cabernet Sauvignon: It's just about perfect. Perhaps a medium-bodied Merlot with a spicy black-pepper character would be a better comparison. However you decide to compare this wine, you'll discover that it's perfect with barbecued meats. This is one of the world's best value wines.

The wine prior to this was the $100 bottle of Clos de Vougeot. Once students discover that this Crozes retails for less than $20, they are ready to buy ten cases!

> Retail Price $20 **BEST VALUE**
> Ready to Drink
> I would also suggest tasting Crozes Hermitage Thalabert from Jaboulet.

Note: The order of the wines should change if you substitute an Hermitage for the Crozes-Hermitage: Taste the Châteauneuf du Pape first since it would be lighter than the Hermitage.

Wine #10: Châteauneuf du Pape, La Bernardine, Chapoutier 2003

This wine is from the southern region of the Rhône Valley and by law is permitted to use thirteen different grape varieties and must have a minimum of 12.5 percent alcohol. Most Châteauneuf du Papes will be 13.5 percent alcohol and higher, so you will often see critics describe it as big, bold, powerful, inky, rich, and robust. The La Bernardine is made from the three best grape varieties—Grenache, Syrah, and Cinsault—but is not as big in style as most Châteauneuf du Papes. This wine displays a balance of fruits, tannins, and alcohol, with less acidity than the Pinot Noirs.

Retail Price $46
Ready to Drink, but can still age
Other recommended producers of "big" style Châteauneuf du Papes are: Beaucastel, Vieux Télégraphe, Clos des Papes, and Rayas. For a lighter style such as the La Bernardine, try the Mont-Redon, Delas Fréres, or Guigal.

Châteauneuf du Pape translates as "new house of the Pope," a name given at the time of the great schism, when the Papacy abandoned Rome and relocated to Avignon at the Southern perimeter of this region.

Buyer beware! It is important that you purchase Châteauneuf du Pape from a reliable producer. There are many variations of style and density, with some as light as a Côte du Rhône, and others big, full-bodied, and luscious. Remember winemakers are allowed to use thirteen different grape varieties blended in any percentage and the best producers use only the best grapes.

CLASS FIVE: RED WINES OF BORDEAUX

We now enter the second half of the wine-tasting course and the ideal opportunity to showcase my favorite red wines: those of Bordeaux. Bordeaux has more than seven thousand châteaux, which makes narrowing down to just ten very difficult.

The three major grapes cultivated in Bordeaux are Merlot, Cabernet Sauvignon, and Cabernet Franc. Merlot is the most widely planted with almost twice as many vines planted as Cabernet Sauvignon, and three times the number of vines as Cabernet Franc.

Bordeaux wines have a reputation of being expensive and requiring long aging. This is not entirely true: 80 percent of all Bordeaux wines retail between $8 and $25 and most of those can be consumed when you buy them or within two years of purchase.

Wines for Class Five:

1. Blind (Appellation Bordeaux Controlée)
2. Barton & Guestier Margaux 2003
3. Château Larose Trintaudon 2001 **BEST VALUE**
4. Château Talbot 2001
5. Clos de Marquis 1999
6. Château Léoville Las Cases 1999 **BEST VALUE**

7. Blind (Cru Bourgeois)
8. Blind (Third, Fourth or Fifth Growth)
9. Blind (First or Second Growth)
10. Gruaud Larose 1989

There are four parts to this tasting.

PART I: UNDERSTANDING BORDEAUX (FOUR WINES POURED TOGETHER, PREFERABLY OF THE SAME VINTAGE)

Wines #1 through #4 represent the major Bordeaux styles, price, and quality levels, and illustrate the wide diversity of Bordeaux red wines. Tasting these four wines together illustrates how the appellation controllée system of Bordeaux works, ranging from the simplest Appellation Bordeaux Controllée to the highest Grand Cru Classé classification.

The lowest appellation of these wines is Appellation Bordeaux Controllée. The B & G Margaux (Appellation Margaux Controllée) is a higher appellation. The next quality level is Château Larose Trintaudon (Château/Haut-Médoc Appellation/Cru Bourgeois), and the highest quality is Château Talbot 2001 (Château/St. Julien Appellation/Grand Cru Classé/Fourth Growth). There are two tasting sections to Part I. The first section compares Wine #1 and Wine #2 and the second compares Wine #3 and Wine #4. Wine #1 and Wine #2 should be from the same vintage, as should Wine #3 and Wine #4. Best would be having all four wines of the same vintage.

SECTION 1: COMPARING WINES #1 AND #2

By tasting Wine #1 and Wine #2 together, students will experience the texture and taste difference between an Appellation Bordeaux Controllée and the higher level Appellation Margaux Controlée. The B & G Margaux has much more fruit, is deeper in color, and has more tannin than Wine #1. These regional wines are some of the best value wines in all of Bordeaux; I wish more were available in the United States. Other good appellations to look for are Médoc, Pauillac, and Saint-Émilion.

Wine #1: Blind (Appellation Bordeaux Controllée)

This Appellation Bordeaux Controlée wine will show a good Bordeaux wine at a great price. It's an easy-drinking, medium-fruit, balanced-tannin, everyday wine that you can buy for tonight's dinner. Most Appellation Bordeaux Controllée wines will be a blend of Merlot, Cabernet Sauvignon, and Cabernet Franc with the primary grape being Merlot. If you're going to buy an Appellation Bordeaux Controllée, look for a great vintage such as 2003 or 2005.

> Retail Price: Most Appellation Bordeaux Controllée wines are under $10
> Ready to Drink
> Other recommended producers are: Mouton Cadet, Michel Lynch, or Lacour Pavillon.

Wine #2: Barton & Guestier Marqaux 2003

The next level of quality in Bordeaux is from one of its fifty-seven different appellations. This wine is from the appellation Margaux, which is a village that produces some of the best wine in Bordeaux. The grapes for this wine can only come from the delimited area of Margaux.

Barton & Guestier is a large négociant who either buys grapes or wine from specific appellations and bottles it under their own name. As with the Appellation Bordeaux Controlée level, I would only buy this wine in a great vintage year such as 2003 or 2005.

> Retail Price $25
> Ready to drink, but could age
> Another recommended négociant is the Baron Philippe de Rothschild S.A.

SECTION 2: COMPARING WINES #3 AND #4

Wine #3 and #4 are both classified Château wines. Wine #3, the Haut Médoc is a lesser classification (Cru Bourgeois), and Wine #4, the St. Julien, carries a much higher classification of Grand Cru Classé. 2001 was a very good vintage year in Bordeaux.

Wine #3: *Château Larose Trintaudon 2001 (Haut Médoc)*

This wine comes from the Haut Médoc appellation and is classified as a Cru Bourgeois. Château Larose Trintaudon is the largest vineyard in the Médoc region. The primary grape of this wine is Cabernet Sauvignon blended with Merlot and Cabernet Franc. It is a medium-bodied Bordeaux with great fruit balance, firm tannic structure, and a moderately long finish.

> Retail Price $20   **BEST VALUE**
> Ready to Drink, but can age
> Other recommended wines for this style are: Château Meyney, Château Les Ormes-de-Pez, Château Greysac, or Château Phélan-Ségur.

Wine #4: *Château Talbot 2001 (St. Julien)*

This wine is from the St. Julien appellation and is classified a Fourth-Growth Bordeaux. I often buy château wine from St. Julien for their subtlety and charm. This is a classic Bordeaux with earthy black fruit, medium to full body, soft tannins, and a long finish. I could "suffer" through this wine tonight but it will be at its best in the next five to ten years. The earliest records of the Château Talbot go back to the seventeenth century.

> Retail Price $45
> Needs aging
> Other recommended wines of this St. Julian style are: Château Gruaud Larose, Château Léoville Barton, Château Branaire-Ducru, Château LaGrange, or Château Ducru Beaucaillou.

PART II: UNDERSTANDING SECOND LABELS (TWO WINES, TASTED TOGETHER, PREFERABLY OF THE SAME VINTAGE)

Second-label wines are usually produced from a château's youngest vines. The wines they produce are lighter in style and quicker to mature than those made from the grapes of older, classified stock. Second-label wines will cost at least one third to a half of that of the château wine and are another way of buying good Bordeaux wine without breaking the budget. Although second-label wines are not a new concept in Bordeaux—the Clos du Marquis produced its first second-label wine in 1902—most châteaux began producing second-label wines in the mid-1980s.

By taking the second-label wine of Château Léoville Las Cases, the Clos du Marquis 1999, and tasting it side by side with Léoville Las Cases 1999, you will see the value of looking for second labels.

Wine #5: Clos du Marquis 1999

All the great châteaux of Bordeaux now produce a second label, and Clos du Marquis is considered one of the best and most consistent in quality. It is by no means a "second" wine, and in great years the Clos du Marquis is of the same or better quality than many of the classified growths. The year 1999 was very good in Bordeaux; this wine, nine years later, has its components in perfect harmony with firm tannins and forward fruit. (See list of second labels on page 131.)

Retail Price $25 **BEST VALUE**
Ready to drink, but will age
Other recommended second-label Bordeaux wines are: Réserve de la Comtesse (Pichon Lalande), Les Tourelles de Pichon (Pichon Longueville), or Réserve du Général (Palmer).

Wine #6: Chateau Léoville Las Cases 1999

One of my favorite vineyards in all of Bordeaux is Château Léoville Las Cases: a perennial favorite and neighbor to my ultimate wine, Château Latour. It is one of the oldest (1638) and largest vineyards in Bordeaux.

This wine usually contains a high percentage of Cabernet Sauvignon, anywhere from 65 to 75 percent. The rest of the blend is made up of Merlot and Cabernet Franc. Even after eight years this wine maintains a dark ruby color with all of the classic Bordeaux superlatives of cedar, vanilla, black cherries, and cassis.

Retail Price $100
Ready to drink, or can age
Other recommended second growth Bordeaux wines are: Château Léoville Barton, Château Pichon Lalande, or Château Pichon Longueville.

PART III: HIERARCHY OF BORDEAUX: THE VALUE OF CRU BOURGEOIS (ALL THREE WINES ARE TASTED BLIND)

Wine #7: Cru Bourgeois

Wine #8: Classified Château (Fourth or Fifth Growth)

Wine #9: Classified Château (First or Second Growth)

This is one of the highlight tastings of my eight-week course. Even though all three wines represent different levels of Bordeaux, tasting them requires subtlety and discernment. They are more difficult to taste because all are of very high quality. These should be tasted blind, which means pour the wines to taste but do not reveal the name of the wine, its vintage, or its cost.

Take at least six minutes to try all three wines without conversation. After recording your thoughts and notes, discuss your results with other participants and take a vote, judging preference, which wine had the most fruit, which wine displayed the most tannin, and judge each wine on whether it's drinkable now or needs more time.

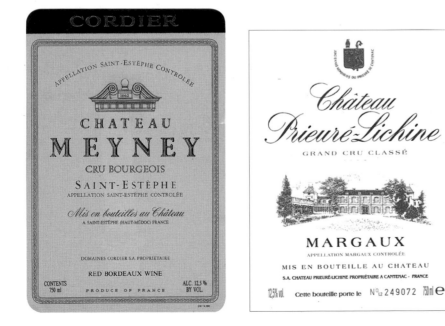

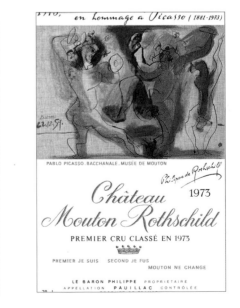

You should find the results quite interesting. There are many times when all three wines receive equal amounts of votes or, depending on the size of your group, you may find that some prefer the first or second growth (Wine #9), some choose the fourth growth (Wine #8), while others select the Cru Bourgeois (Wine #7). Ask yourselves if the wines are from the same château, three different vintages or are they from different châteaux, all the same vintage?

The point is not which wine wins the popular vote but how closely all three are judged. Now reveal the château, vintage, and price; everyone will learn an important lesson in the true value of the lesser appellations. The Cru Bourgeois is $25, the third, fourth, or fifth growths are $50 to $75, and the first or second growths are $100 to $400 per bottle. This reinforces the message to my students that you don't have to spend a lot of money to buy a good bottle of Bordeaux.

PART IV: ENJOYING AN AGED BORDEAUX (ONE WINE, TASTED ALONE)

Wine #10: Gruaud Larose 1989 (St. Julien)

This is a second-growth Bordeaux from the village of St. Julien. Gruaud Larose is one of the most well-known Bordeaux châteaux. This property was created by the Gruaud brothers in the mid-1700s and today is one of the "super seconds"—or best of the second-growth Bordeaux châteaux-wines of the 1855 Classification. The primary grape is Cabernet Sauvignon (around 60 percent) blended with 25 to 30 percent Merlot, with the rest Cabernet Franc and Petit Verdot.

Pay close attention to how aging affects a Bordeaux wine. The color is lighter, there are fewer tannins, the fruit is more pronounced, and it has an intoxicatingly different bouquet and aroma than younger Bordeaux. There is a slight bit of herb, tobacco, and an autumnal "dead leaf" bouquet. This wine has reached its peak and needs to be consumed without further aging. Ask yourself what you were doing in 1989 and if you have aged as well as this wine!!

Retail Price $100+
Ready to Drink
It's unlikely that you will be able to locate an eighteen-year-old Bordeaux, but the point of this tasting is to notice how a Bordeaux is affected by age, so try your best to find an older vintage.

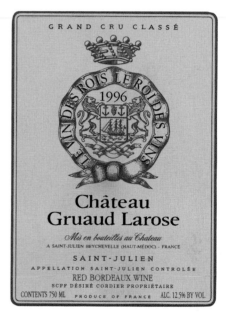

Wines are my props. The order of the wine gives structure and purpose to the lecture, allowing me to discuss the appellation system, grape varieties, vinification, and ageability of Bordeaux wines.

CLASS SIX: RED WINES OF CALIFORNIA, OREGON, AND WASHINGTON

This is the sixth class and you've tasted more than fifty wines. You've ingested a tremendous amount of information—and wine!—over the last five classes; perhaps too much. So now we relax. Class Six is the "break" class in which we veer away from too many facts and statistics and where I join the class in each 60-second wine expert tasting. I give my personal opinion of each wine as I taste it. It's also the only time in the course that the students observe how a professional wine taster judges wine! This is the class after which I once received the best compliment ever: A student told me that this was the best class of all because she now understood which wines she liked and why. Intrigued, I asked her what about this class was so invaluable. Her answer surprised and delighted me—"Every wine you liked, I hated!" Clearly we were incompatible in our tastes, but she was honest and had discovered more about her own tastes—that's the nature and purpose of wine tasting.

In Class Two we covered the white wines of California and the United States. In this class I hope you'll learn more about tasting, and further identify your own preferences on wine styles. While I can't be with you as you taste, the notes that follow each tasting are my own.

Wines for Class Six

I am using a Pinot Noir from Oregon and a Merlot from Washington State. These two states are increasingly recognized as producing some of the best wines in the United States.

1. Saintsbury Pinot Noir 2004
2. Domaine Drouhin Pinot Noir (Oregon) 2004
3. Ravenswood Zinfandel Sonoma Series 2004
4. Ridge Geyersville Zinfandel 2003
5. Columbia Crest Merlot 2003 (Washington) **BEST VALUE**
6. Phelps Merlot 2003
7. Louis Martini Cabernet Sauvignon 2003 Napa **BEST VALUE**
8. Blind
9. Blind
10. Blind
11. Robert Mondavi Reserve Cabernet Sauvignon 1996

The first six wines are all poured together so that you can clearly see the difference in color between the Pinot Noirs, Zinfandels, and Merlots. The Pinot Noirs are much lighter in color than the Zinfandels and Merlots, and most students judge the Merlot-based wines as the deepest in color. I hope that after you've tasted these first six wines, you will be able to choose the grape variety that best suits your personality: your "eureka" moment.

There are six parts to this tasting.

PART I: PINOT NOIR (TWO WINES TASTED TOGETHER FROM THE SAME VINTAGE)

Within wine circles there is always a heated debate over which state produces the best Pinot Noir: California or Oregon. My answer is always the same: Both states produce some of the best Pinot Noir in the world, but each state produces a different style.

Wine #1: Saintsbury Pinot Noir 2004

The Saintsbury Winery of Carneros, California, was named after the British wine writer George Saintsbury and founded in 1981. Owners Dick Ward and David Graves pioneered growing Chardonnay and Pinot Noir grapes and are well known today for the many different styles they produce of each.

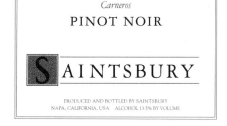

I've used the Saintsbury Pinot Noir many times in my class because of its consistent high quality and reasonable price. This 2004 Pinot Noir is medium in body with a classic Pinot Noir bouquet and hints of spice and black cherry. The complexity of the taste lingers, giving this wine a soft and elegant finish. Although this wine has been aged in small French oak barrels, the oak never dominates the fruit.

Retail Price $33
Ready to Drink
Other recommended producers are: Acacia, Etude, or Flowers.

Wine #2: Domaine Drouhin Pinot Noir 2004

I chose the Domaine Drouhin Pinot Noir because of the family's hundreds of years of winemaking expertise in Burgundy, France, where the ultimate expression of Pinot Noir developed. Robert Drouhin visited the Willamette

Valley back in the late 1960s and, in 1989, opened Domaine Drouhin. This 2004 Pinot Noir (a great year) is a classic-style Pinot Noir, delicate yet complex, bigger in body than the Saintsbury with lots of red fruit in the smell and taste. What's the bottom line? This is a delicious wine.

Retail Price $40
Ready to drink
Other recommended producers are: Eyrie, Archery Summit, Beaux Frères, or Rex Hill.

PART II: ZINFANDEL (TWO WINES TASTED TOGETHER, PREFERABLY FROM THE SAME VINTAGE OR ONE YEAR APART)

The red Zinfandel grape variety has been part of California's history for more than a hundred years. It was a favorite of Italian winemakers since it was reminiscent of the many Italian wines. The best Zinfandel will produce big, full-bodied, robust, high-alcohol, long-lived wines. If you are looking for this style of Zinfandel, be aware that many of the best wineries also produce a very light- to medium-style wine. Ravenswood and Ridge are considered to be two of the top Zinfandel producers in California.

Wine #3: Ravenswood Zinfandel Sonoma Series 2003

Joel Peterson, the winemaker and cofounder of Ravenswood, makes his wines the way he lives his life: no wimpy wines. You can be assured that you'll always get a high-quality wine at every price level, whether it's his Vintner's Reserve at $12 or his vineyard designated Pickberry at $54. The 2003 Ravenswood Zinfandel Sonoma County is a blend of 95 percent Zinfandel, 4 percent Carignane, and 1 percent "mystery" grapes. It spends twenty-four months in French oak 35 percent new, is typical of the Zinfandel grape in its 14.5 percent alcohol level, and has plenty of fruit and tannin with undertones of raspberry and blackberry flavors.

Retail Price $18
Ready to Drink
Other recommended wines for this style are: St. Francis, Rosenblum, Rochioli, and Dry Creek.

Wine #4: *Ridge Geyserville Zinfandel 2003*

The ultimate Zinfandel from California is made by Paul Draper, who has been a winemaker for Ridge since 1969. Although Ridge makes one of the best Cabernet Sauvignons in California, Zinfandel was their first specialty. Ridge produces many Zinfandels from different vineyards throughout California and in the 2003 vintage they offered seven vineyard designated Zinfandels. The wine you are tasting comes from the Geyserville vineyard in Sonoma County. A blend of 76 percent Zinfandel, 18 percent Carignane, and 6 percent Petite Sirah, This Zinfandel is 14.6 percent alcohol. It is a big, full-bodied, chewy, earthy, jammy wine that still maintains the elegance of the Zinfandel fruit.

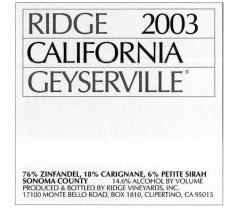

Retail Price $35
Needs more time
Other recommended wines of this style are: Turley or Seghesio.

PART III: MERLOT (TWO WINES TASTED TOGETHER FROM THE SAME VINTAGE)

Ever since the movie *Sideways* came out, people have become insecure about drinking Merlot. Don't be. It is one of the best red grapes grown in California and Washington State and, of course, as we tasted in Class Five, some of the best wines of Bordeaux, France, are made from the Merlot grape. With all of that said, the consumer should not take the quality of Merlot for granted. It is extremely important to know the best producers.

Wine #5: *Columbia Crest Grand Estates Merlot 2003*

The first vintage of Columbia Crest was 1984, and four years later in 1988 they released their first Merlot. Since that time, Columbia Crest Merlot has been listed in my book as one of the best-quality red wine values produced in the United States.

This wine comes from the Columbia Valley and is made from Merlot (more than 90 percent) and a blend of Cabernet Sauvignon and Cabernet Franc. It has 13.5 percent alcohol content, is medium-bodied, and has plenty of accessible fruit with well-balanced tannins. The wine was fermented in stainless steel vats, underwent malolactic fermentation, and was aged for four-teen months, some in French oak and some in American oak.

Retail Price $16   **BEST VALUE**
Ready to Drink
Another recommended Washington State value Merlot producer is
Hogue Merlot.

Wine #6: Phelps Merlot 2003

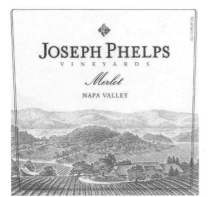

Joe Phelps has always been one of the most upstanding citizens of the Napa
Valley. He was a building contractor from Colorado who moved to the Napa
Valley in the early seventies. After building a few other wineries, he built his
own, the Joseph Phelps Winery, in 1973. Phelps has only had two winemakers;
Craig Williams, who joined Phelps in 1983, continues to produce consistent,
high-quality wines. This Phelps Merlot is made from 100 percent Napa Valley
grapes fermented in stainless steel tanks before aging in a combination of
new and older French oak barrels. This is a full-bodied, intense, rich wine
with hints of chocolate and cinnamon.

Retail Price $36
Ready to Drink, and can be aged
Other recommended Napa Merlot wines are: Markham, Shafer, or
Whitehall Lane.

The next five wines are all made with Cabernet Sauvignon.

PART IV: INTRODUCTION TO CABERNET SAUVIGNON (ONE WINE, TASTED ALONE)

My favorite red grape is Cabernet Sauvignon. It is also one of the best grapes
grown in California, particularly in the Napa Valley. The diversity of style and
price of California Cabernet Sauvignons is abundant. Are you looking to
spend $15 or $150 or $550? There's something for everyone and any occasion.

Wine #7: Louis Martini Cabernet Sauvignon 2003

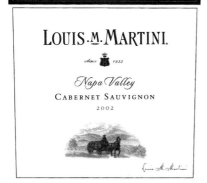

I've had sentimental feelings for the Louis Martini Winery ever since I created
my first wine list for the Windows on the World restaurant. Louis Martini
Cabernet Sauvignons were featured on every one of my lists for the twenty-
five years I worked there as wine director. The Martini family is one of the
oldest families of quality winemaking in California. They started their winery

in 1922 and moved to the Napa Valley right after prohibition ended. They have always created good quality, primarily red, honest wines at affordable prices. I have been fortunate to have tried many of their early Cabernet Sauvignons going back to the 1940s.

All the grapes for this 2003 Cabernet Sauvignon come from the Napa Valley. The two primary grapes are Cabernet Sauvignon (78 percent) and Merlot (18 percent). Eighty-five percent of the wine was aged in oak barrels for eleven months. The wine has an alcohol content of 14.2 percent.

This is not a blockbuster Cabernet Sauvignon, which is the main reason that I use it as my first Cabernet. For as long as I have known the wine, it has been medium in body with really good flavor and structure. It's a great restaurant wine and today remains one of the best values of all of California Cabernet Sauvignons.

Retail Price $22 **BEST VALUE**
Ready to Drink, but can age
Other recommended producers of this style Cabernet Sauvignon are:
Beringer Founders Estate, Simi, or Gallo of Sonoma.

PART V: BLIND (THREE DIFFERENT CABERNET SAUVIGNONS)

Wine #8: Blind

Wine #9: Blind

Wine #10: Blind

This is a very special tasting and much like the tasting we did with the wines of Bordeaux in Class Five. The only thing I let the students know is that these wines are three high-quality Cabernet Sauvignons coming out of my personal wine cellar. The students first discuss the color and the smell, just as they did in the previous class. They now have six minutes in silence to try the three wines. I do this tasting blind because it helps the students concentrate more on the individual characteristics of each wine. Somehow when you see a label, you get a preconceived impression of what the wine should taste like. I also ask the same question as I have in Class Five: Are these Cabernets all from the same vintage or different vintages? See page 138 for a listing of my favorite Cabernets.

Because my students have to read this book for my class, I cannot share the three wines I serve with you. Just use your imagination and think high quality.

PART VI: CABERNET SAUVIGNON RESERVE (ONE WINE, TASTED ALONE)

Wine #11: Robert Mondavi Reserve 1996

I would not be writing this chapter, would not have been able to write my wine book, and probably wouldn't have been able to operate my wine school for more than thirty years without the vision of Robert Mondavi. Mondavi set the standards for quality California wine long before anyone was interested, and his insights traveled well beyond the borders of Napa Valley, California, and the United States. He influenced winemakers throughout the world.

There is no easier way to explain or describe what happens to wine as it gets older or what it tastes like than serving an older Cabernet Sauvignon and allowing you to formulate your own opinion. This 1996 Robert Mondavi Cabernet Sauvignon Reserve is primarily made with Cabernet Sauvignon and most of the grapes come from its To-Kalon Vineyard, arguably the best vineyard in Napa. At eleven years old, this wine still has a deep color with earth tones of minerals, and intense flavors of cassis and black fruit. This wine is still in its peak period. It can be consumed now or continue to age.

> Retail Price $175
> Ready to Drink, but can age
> Other recommended producers of Napa Valley reserve wines are:
> Beaulieu Private Reserve, Beringer Private Reserve, Shafer Hillside Select, Silver Oak, or Stag's Leap Cask 23.

CLASS SEVEN: RED WINES OF SPAIN AND ITALY

I look forward to this class since many of my students know something about French or American wines but very little about Italian or Spanish wines. This is a class of discovery: finding new wine styles made from indigenous grapes such as Nebbiolo, Sangiovese, and Tempranillo. The wines of both countries have improved dramatically over the last twenty years.

Wines for Class Seven: Red Wines of Spain and Italy

1. Cune Vina Real Crianza 2005 **BEST VALUE**
2. Conde de Valdemar Reserva 2001 **BEST VALUE**
3. La Rioja Alta Gran Reserva 1995
4. Chianti Ruffina Riserva 2003, Nipozzano
5. Vino Nobile di Montelpulciano, 2001, Avignonesi
6. Brunello di Montalcino 1997, Banfi
7. Barbera, Le Orme 2004, Michele Chiarlo **BEST VALUE**
8. Barbaresco 2001, Vietti
9. Barolo 2000, Prunotto
10. Amarone 2000, Allegrini

There are four parts to this tasting.

PART I: RIOJA (THREE WINES TASTED TOGETHER)

Historically, Rioja has been the center of quality winemaking in Spain, and is now producing some of its best wines ever. Not only are the larger, established producers making better wines, but Rioja has enjoyed a surge in small artisan producers as well. The primary grape variety for all of the Rioja wines is the Tempranillo grape.

These three wines are poured together to show the three different quality levels of Rioja. You'll taste in order from the lightest Crianza, to a bigger Reserva, to the best quality level Gran Reserva.

Wine #1: C.U.N.E. Vina Real Crianza 2005

C.U.N.E. is one of the most recognized Bodegas in the United States, producing a consistent quality wine at all levels. I have chosen their Crianza, which goes under the proprietary name Vina Real. This wine is a blend of 90 percent Tempranillo and 10 percent Garnacha (same grape that grown in the Rhône Valley is called Grenache). It is an easy-drinking, fruit-friendly wine with balanced soft tannins. Crianza goes well with all kinds of food and is delicious all by itself. All Crianza should be consumed within five years of the vintage.

Retail Price $15 **BEST VALUE**
Ready to Drink
Other recommended Crianza-style producers are: Bodegas Montecillo,
Marques de Caceres, or Marques de Riscal.

Wine #2: *Conde de Valdemar Reserva 2001*

The original winery was established in 1889 and it is owned by the Martinez Bujanda family. It was one of the first wineries in Rioja to make a more modern-style wine (less oak aging with more accessible fruit when young). The Reserva is made from 85 percent Tempranillo and 15 percent Mazuelo and is aged twenty-five months in a combination of French and American oak. This wine is medium-bodied, has tremendous fruit character and great acidity that lingers into the aftertaste with tannin as the backbone. Rioja had a great vintage in 2001.

Retail Price $17 **BEST VALUE**
Ready to drink
Other recommended producers of Reserva wines are: Bodega Muga,
C.U.N.E., or La Rioja Alta.

Wine #3: *La Rioja Alta Gran Reserva 1995*

A Gran Reserva is usually released after five to seven years of aging with a minimum of two years in oak barrels, and 1995 was a great vintage in Rioja. Many wine critics feel that La Rioja Alta is one of the great international wineries and I agree with them. For over one hundred years they have been producing high-quality wine. This Gran Reserva is 85 percent Tempranillo blended with Graciano and Mazuelo grapes. The 1995 vintage shows the finesse and intensity of a great aged Tempranillo wine, full-bodied, mouth-filling, and spicy with great acidity and balanced tannins, and it will continue to improve with age.

Retail Price $52
Needs to age longer
Other recommended producers of Gran Reservas are: C.U.N.E. or
Bodegas Muga.

Part II: Tuscany (three wines tasted together)

If I had to move today and was given a choice of living anywhere in the world I would want to live in Tuscany, Italy, not just because they have some of my favorite wines, but because of its people, mountains, proximity to water, and the abundance of superb food. I first visited Tuscany when I was twenty-four years old when most wineries were making decent, though ordinary, wine. Those were the days when olive trees were interspersed between the vines. Thirty years later, Tuscany has gone through one of the biggest transformations of any major wine region in the world. Through modern technology, the wineries of Tuscany are now producing more and better wines.

Although there is no official hierarchy that defines the quality of Tuscan wine, these three wines should be tasted in the order that I've placed them. You'll find that the Chianti is much lighter than the Vino Nobile di Montelpulciano, and the Vino Nobile di Montelpulciano is lighter than the Brunello di Montalcino.

Wine #4: Chianti Rufina Riserva 2003, Nipozzano

The Nipozzano Chianti Rufina Riserva is made by the Frescobaldis, a seven-hundred-year-old Tuscan family. This is a medium-style wine primarily made from the Sangiovese grape (90 percent). Tuscany had a great vintage year in 2003, and this wine, after five years, is in perfect condition to drink. There is complexity in the fragrance of black currant and spicy aromas. Sangiovese is known for its high acidity. It is a perfect complement to a Sunday Italian dinner with pasta, gravy, and meatballs.

> Retail Price $24
> Ready to Drink
> Other recommended Chianti producers are: Antinori, Ruffino, Brolio, or Melini.

Wine #5: Vino Nobile di Montelpulciano 2001, Avignonesi

Avignonesi is one of the best producers of Vino Nobile di Montelpulciano, which means "noble wine from the village of Montelpulciano." The vineyard has a long history going back to the late 1300s, and since 1974 has been owned by the same family. The primary grape is Sangiovese. The wine was aged in oak for twenty-four months. Garnet-red in color with a soft, delicate bouquet, the wine still has more tannins and higher acids than fruit. This wine will need age in order to soften the tannins or, if you can't wait, serve it with cheese or grilled meat.

Retail Price $28
Ready to Drink
Other recommended Vino Nobile di Montelpulciano producers are: Boscarelli or Fassati.

Wine #6: Brunello di Montalcino 1997, Banfi

Castello Banfi owns over 2,400 acres of vineyards in Tuscany producing all different styles of wine, but they put their best effort into Brunello. This is the biggest and fullest in body of the three Tuscan wines. Eleven years later this Brunello still tastes youthful, full of fresh 100 percent Sangiovese fruit. Some Tuscan winemakers feel that the 1997 vintage was the best since 1947. With two-and-a-half years of barrel aging and 13 percent-plus alcohol, this wine still needs another five years or more before it reaches its peak.

Retail Price $75-$100
Needs more time
Other recommended Brunelo di Montalcino producers are: Barbi, Poggio Antico, or Col d'Orcia.

PART III: PIEDMONT (THREE WINES TASTED TOGETHER)

The taste and style of Piedmont wines are quite different than those of Tuscany. Where Sangiovese is the major grape in Tuscany, the Nebbiolo grape is the king of Piedmont. Piedmont wines, especially in the villages of Barolo and Barbaresco, are bigger, fuller in body, and higher in alcohol than those of Tuscany. These are wines that you need to cellar for many years before they will be ready for consumption. In my class I advise my students that when dining out in restaurants they should buy Tuscan wines: they will be more accessible when young, and are lower in price than the wines of Piedmont. Barolos and Barbarescos need ten to fifteen years of age. Age them in your home wine cellar and serve them with a great homemade Italian meal.

Wine #7: Barbera Le Orme 2004, Michele Chiarlo

Barbera is one of the most widely planted grape varieties in Italy and comprises 50 percent of the vines planted in Piedmont. This 100 percent Barbera comes from the village of Asti and is made by one of the best Piedmont producers, Michel Chiarlo. I like to use this wine as a prelude to my next two because it typifies this region's wine style: medium to full in body, very dry, and robust. This is the wine that you drink now while you wait for your Barolos and Barbarescos to age. It's the everyday drinking wine for the local Piedmontese.

Retail Price $12 **BEST VALUE**
Ready to Drink
Other recommended Barbera producers are: A. Conterno and Prunotto.

Wine #8: Barbaresco 2001, Vietti

Vietti is one of the most consistent and reliable producers in Piedmont. The village of Barbaresco produces wines from 100 percent Nebbiolo grapes. The older style of Barbarescos undergo a longer maceration of the skins and juice, are kept in oak longer, and have a rustic taste with lots of tannins that need years of aging to soften. The Barbaresco Vietti emphasizes the Nebbiolo fruit with fewer tannins and is ready to drink sooner. Piedmont had a great vintage in 2001, and this wine is over 13.5 percent alcohol. It's a big, full-bodied, mouth-filling wine with black fruits and integrated tannins which work better with veal chops than veal scaloppini.

Retail Price $80
Drinkable now, but better with age
Other recommended Barbaresco producers are: Gaja, Michel Chiarlo,
and the Cooperative of Produttori d'Barbaresco.

Wine #9: Barolo 2000, Prunotto

This small little village in Piedmont makes one of the most famous wines in all of Italy. Some of the oldest vineyards in the world are located in Barolo. Prunotto is one of the top producers, and 2000 was a great vintage.

With 14 percent alcohol, the bouquet and taste of this wine are very powerful. On the palate there is a tremendous concentration of the Nebbiolo grape and intense tannins. Right now it is a little austere—this wine still needs at least another ten years of aging to show its true character.

Retail Price $60
Needs aging
Other recommended Barolo producers are: Pio Cesare, Ratti, Conterno, and Marchese di Barolo.

PART IV: VENETO: THE FINAL TREAT: AMARONE (ONE WINE, TASTED ALONE)

Wine #10: Amarone 2000, Allegrini

This wine goes into the "unique" category. Coming from the Valpolicella region in Veneto, it is made primarily with the Corvina variety of grape. The grapes are left on the vines until most of the water has dissipated and the grapes shrivel. After picking the grapes they are then left to dry for several months, concentrating the sugars. Through fermentation this sugar gives the Amarone very high alcohol content.

You will see by its color that this wine is not ready to drink. Even though it is almost seven years old, it is deep, dark, rich in color, and, like the Barolo, needs many more years of aging. I emphasize that it is very important to smell this wine before trying it and ask yourself: what other wines does it remind you of? Often the answer is Port because of its high alcohol (14.5 percent and higher) and its sweet smell. This is the biggest in style of all of the wines we taste during the eight-week course.

Allegrini is one of my favorite producers of Amarone. When it comes to pairing this wine with food, it is too powerful even for most meats. At home I serve this wine with my cheese course.

Retail Price $80
Needs aging
Other recommended Amarone producers are: Masi, Bertaini, Quintarelli, and Tommasi.

CLASS EIGHT: FINAL EXAM, CHAMPAGNE, AND PORT

This is our eighth and last class. It's your graduation class and there's no better way to celebrate than by beginning the evening with Champagne and ending the night with Port.

Normally, I begin this class with a final exam during which the students test their olfactory abilities with a blind tasting of four wines, all of which they have had during the semester.

Wines for Class Eight: Champagne and Port

1. Cristalino Cava Brut NV **BEST VALUE**
2. Roederer Estate Brut NV
3. Domaine Carneros Brut 2004
4. Taittinger Brut La Française NV
5. Veuve Clicquot Brut Yellow Label NV
6. Blind
7. Taylor Fladgate Ruby Port **BEST VALUE**
8. Churchill's Tawny Port – 10 Year Old
9. Ramos Pinto Vintage Port 1994

There are two parts to this tasting.

PART I: CHAMPAGNES AND SPARKLING WINES

Part I is divided into three sections: the sparkling wines of Spain and the United States, and the champagnes of France. Most of my readers want to know the best value choices in wine, especially when it comes to expensive wine alternatives. Although French Champagne remains one of the great value wines in the world a non-vintage bottle still costs around $40 a bottle.

SECTION I: SPARKLING WINE: SPAIN (ONE WINE, TASTED ALONE)

Spain produces some great sparkling wines at unbelievable prices. European Union laws prevent any sparkling wine produced outside of Champagne,

France, to be called Champagne. In Germany it is called Sekt, in Italy it is called Spumante, and in Spain it is called Cava.

Wine #1: Cristalino Cava Brut NV

For many years the #1 sparkling wine in my "value" tastings has been Cristalino. It comes from Catalonia, located in northeastern Spain, which is considered to be the best Cava-producing region in the country. It is made with indigenous grapes produced in the same process as French Champagne (*méthode Champenoise*), and is aged for nearly two years in oak. It is great for weddings, special events, and large parties because of its quality and price. This wine has great effervescence (small bubbles) and is light, refreshing, and easy to drink.

> Retail Price $10 **BEST VALUE**
> Ready to Drink
> Other recommended Cava producers are: Codorniu or Freixenet.

SECTION II: SPARKLING WINES OF CALIFORNIA (TWO WINES TASTED TOGETHER)

California has been making sparkling wines since the end of Prohibition, but didn't produce a quality sparkling wine until the 1965 introduction of a sparkling wine from the Napa Valley winery Schramsberg. In 1973 the French Champagne producer Moët & Chandon created a Napa Valley sparkling wine, called Domaine Chandon, on a larger scale, and the quality of California sparkling wines has been soaring ever since.

It was difficult to emulate a great French Champagne in California because the climatic conditions of sunny Napa were quite different than the colder climatic conditions of Champagne, France. In all honesty, I did not recommend most California sparkling wines until about fifteen years ago. By then many other sparkling wine producers had joined Domaine Chandon and Schramsberg and, through viticultural research, the major quality grapes Chardonnay and Pinot Noir are now planted in the right soil and climate to make excellent sparkling wines.

Wine #2: Roederer Estate Brut NV

Many of my readers are familiar with the French Champagne Roederer. Roederer has been involved in fine winemaking for more than two hundred years.

They chose to plant their North American vineyards and build their winery in the cooler climate of the Anderson Valley in Mendocino County. The Roederer Estate Brut NV is produced exactly as done in Champagne and is a blend of 60 percent Chardonnay and 40 percent Pinot Noir. All of the grapes used to make Roederer Estate Brut are grown by Roederer in their own vineyards, hence the word "Estate" on the label. The wine is fermented in stainless steel tanks with some oak aged reserved wines added to the blend to give the wine more intensity and flavor. The winemaker has left some residual sugar in the wine (1.2 percent), which balances out the acidity and combined with the fruit flavors makes this wine very accessible now.

Retail Price $25
Ready to Drink

Wine #3: Domaine Carneros Brut 2004

This sparkling wine producer was created by another great French Champagne house, Taittinger. I taste the Domaine Carneros after the Roederer for two reasons: This is a vintage sparkling wine; and it contains more red grapes (61 percent Pinot Noir), which gives it more body. All the grapes for this wine come from the Carneros district, the coolest region of Napa/Sonoma, perfect for growing the finicky Pinot Noir. It has also been aged for three years in the bottle and the fruit's acid and carbon dioxide have blended together, making for a clean, crisp, and elegant sparkling wine.

Retail Price $25
Ready to Drink
Other recommended California Sparkling Wine producers are:
Domaine Chandon, Piper Sonoma, Mumm Cuvée Napa, Schramsberg, and Iron Horse.

SECTION III: FRENCH CHAMPAGNES (THREE WINES TASTED TOGETHER)

My favorite sparkling wine in the world is from Champagne, France. I always serve Champagne as an aperitif at dinner parties. The Champagne region has a long history and tradition of fine winemaking going back hundreds of years.

All three wines are poured together. This is the final blind tasting of the course and we will compare two French Champagnes: Taittinger (Wine #4) and

Veuve Clicquot (Wine #5) against a sparkling wine (Wine #6) which could be a French Champagne, an Italian Spumante, or an American sparkling wine.

Wine #4: Taittinger Brut La Francaise NV

The Champagne House Taittinger was founded in 1734 and has produced quality wine for nearly 275 years. I've placed the Taittinger first in this flight for several reasons. It comes right after the Domaine Carneros from California, also made by Taittinger, and you'll experience the differences between their California and French versions. It is also lighter and more delicate than the next wine, the Veuve Clicquot, due to its higher proportion of Chardonnay grapes.

All French Champagnes begin as white wine before undergoing a secondary fermentation in the bottle. By French law, a non-vintage Champagne, which is a blend of different vintages, is required to be aged for a minimum of one year. The Taittinger is aged three to four years during which time it acquires more complexity. The bouquet of this wine is toasty, yeasty, and has a lot of Chardonnay fruit flavors. The fruit, acid, and carbon dioxide are perfectly balanced, giving this wine a very long finish.

Retail Price $36
Ready to Drink

Wine #5: Veuve Clicquot Brut Yellow Label NV

The twenty-fifth anniversary celebration of Windows on the World was planned for October 2001 with the opening of its new wine cellar. The 500 cases of twenty-fifth anniversary Veuve Clicquot Cuvée had not been delivered to the restaurant when it was destroyed on September 11, 2001. The 500 cases were sold to our club members and the proceeds were donated to Windows of Hope, a charitable foundation established for the families of Windows on the World staff who lost their lives.

Clicquot was founded in 1772, thirty-eight years after Taittinger. Veuve Clicquot NV is a sentimental favorite of mine. It was the house Champagne at Windows on the World and I still have bottles from the tenth, fifteenth, and twentieth anniversaries of the restaurant in my cellar.

The Veuve Clicquot Brut Yellow Label NV is a blend of 50 percent Pinot Noir, 35 percent Chardonnay, and 15 percent Pinot Meunier. The Pinot Noir gives this wine more body, the Chardonnay freshness, and the Pinot Meunier fruitiness. It is a big, full, rich style Champagne and the effervescent tiny bubbles enhance a bouquet of toasty Brioche, fruit, and apple aromas resulting from four years of aging.

Retail Price $40
Ready to Drink

See page 180 for other recommended French Champagne producers.

Wine #6: Blind

As a buyer of wine for thirty years, I have found the only true way to understand the quality of a wine, especially Champagne, is to compare it to well-established and universally-acknowledged standard bearers. Can Wine #6 live up to the high standards of the two Brut non-vintage Champagnes, Taittinger and Veuve Clicquot? Choose your own "mystery" wine for this blind tasting and pour it for your friends. Once they've tasted all three wines, ask them if Wine #6 costs the same, more, or less than the Taittinger and Clicquot. Is it a French Champagne, an Italian Spumante, or an American sparkling wine? Some people may think it's an inexpensive California sparkling wine and others might believe it to be a Dom Perignon!

Part 11: Port (three wines tasted together)

Port is a cool-weather wine; with 20 percent alcohol, it warms the body quickly! I only drink Port between October 15 and April 15 and then only if the temperature is 20 degrees or lower. Ideal conditions include a late evening snowfall with all my children snuggled in their beds, a crackling wood fire in front of me, and my golden retriever by my side! My only exception to this rule is when a class falls before or after these dates. Fortunately, my spring class usually ends before April 15 and my fall class ends around November 15.

There are three different quality levels to this tasting and the wines must go in the following order: first Ruby, then Tawny, and finally the Vintage Port.

Wine #7: Taylor Fladgate Special Ruby Port

Taylor Fladgate, founded in 1692, is one of the world's great Port producers. Both their Ruby Port and their Tawny Port are cask-aged ports—aged in the cask for at least three years, much longer than a Vintage Port. Most Ruby Ports should be consumed immediately after you buy them and will not age in the bottle. The Taylor Fladgate Ruby Port is deep and dark in color, fruity in the nose, and easy to consume with a great balance of fruit, sweetness, and low tannins. With the high alcohol and residual sugar (6 percent), Port is not for everyone.

Retail Price $15 **BEST VALUE**
Ready to Drink

See page 188 for other recommended Ruby Port producers.

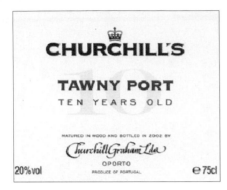

Wine #8: Churchill's Tawny 10 Year Port

Just looking at the color of this wine will help you understand why it's named Tawny. This wine has been aged for ten years, and red wines lose color with age. You can see through this wine, which has a tawny garnet, pale color with a slight brown tinge. It is known as an "all year" Port because of its lighter style. Twenty-, thirty-, and forty–year-old Tawny Ports are also available but obviously, the longer the aging, the higher the price.

With its nutty bouquet and high acid content the Churchill's 10 Year Old Tawny is the wine I serve after my Thanksgiving meal with walnuts, pecans, and fruit and is a great example of what a Tawny Port is all about.

Retail Price $30
Ready to Drink

Wine #9: Ramos Pinto 1994 Vintage Port

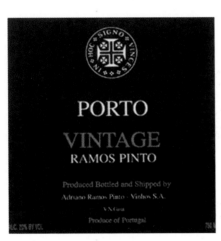

Port can only be classified a vintage if all of the grapes used in its making are harvested in the same year. Some critics believe that the 1994 vintage is the greatest vintage since 1945. Whereas Ruby and Tawny ports might age in cask for any number of years, Vintage Port is aged for only two years, after which it continues to age in the bottle for another ten to one hundred years, depending on the vintage. Of the three Ports this is the only wine that contains sediment. All Vintage Ports should be decanted before serving, especially as they get older.

Vintage Port is a truly exceptional wine and is, along with French Bordeaux and California Cabernet Sauvignons, one of the great collectibles.

The initial taste of the Ramos Pinto is sweetness, but thirty seconds later the tannins are evident. So what you have is high tannin, alcohol, and residual sugar creating a bittersweet taste in this wine. Some students love it, others don't. With time, all of the components will mellow into a great wine experience. This 1994 with fourteen years of age on it is still not ready to drink, and probably won't be, for at least another ten years.

Retail Price $100
Needs to Age: 10 years

Other recommended Vintage Port producers and vintages are on page 188.

Continuing Your Tasting Experience

YOUR EXPERIENCE with wine never truly ends. But there is only so much I can cover, even in an eight-week course, so I choose to focus on the regions and grapes that represent a large selection of the wines you are likely to encounter, as well as offer the best opportunities for teaching about the attributes that contribute to the distinct experience of each type of wine and each individual wine.

I encourage my students to seek out wines from regions we don't cover in the class, and wines made with grape varieties we don't have the time to taste during our time together. The world of wine is much bigger than France and the United States! There are specific types of wines that I particularly think you must taste in order to have a fuller appreciation of wine. These are listed on the next page.

How you go about it is up to you—think of this as the essay question in your wine-tasting education! You may want to taste a selection of, for example, Argentine Malbecs together, comparing them much as we did for the other wines in this tasting. Or you may decide to try one wine from the list and compare it with similar wines from another region—maybe you'd like to see what distinguishes a Cabernet Sauvignon made in Italy from a California Cabernet Sauvignon by tasting them together. In general, I recommend tasting wines from lighter to heavier (or else the heavier wine will "drown out" the sensations produced by the lighter wine), but other than that, experiment. Use this as your opportunity to develop your own methods and preferences for tasting wines.

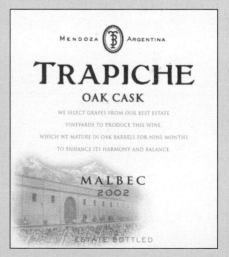

Try some (or all) of these:

Cabernet Sauvignon from Chile
Malbec from Argentina
Sauvignon Blanc from New Zealand
Shiraz from Australia
Pinotage from South Africa
Gruner Veltliner from Austria
Pinot Grigio from Italy
Cabernet Sauvignon from Penedes, Spain
Moschofilero from Greece
Ice Wine from Canada
Kosher Wine from Israel
Portuguese Red Table Wine
American wine from states beyond California, Washington,
    Oregon, or New York*
Vinho Verde from Portugal
Manzanilla Sherry
Vouvray from Loire Valley
Rainwater Madeira
Valpolicella Superiore Ripasso

I hope this wine-tasting course has given you the foundation on which to build a lifetime of experiencing wine—the knowledge and vocabulary to appreciate the richness and subtleties of the wines you drink, to know what you like and don't like and why, and the ability to articulate and share your wine experience with others. Enjoy!

*Fine wines aren't restricted to those four states: viniculture and winemaking is a growth industry throughout the United States. If you'd like suggestions for other American wines to try, see *Kevin Zraly's American Wine Guide*, in which I have included wines and wineries from all fifty states. You may be surprised to find a fine wine close to home—wherever your home may be.

The Physiology of Tasting Wine
You Smell More Than You Think!

BY KEVIN ZRALY AND WENDY DUBIT

WINE SMELLING 101

ONE OF THE MOST wonderful things about wine is its ability to bring us to our senses. While all of our senses factor into the enjoyment of wine, none does so powerfully or pleasurably as olfaction, our sense of smell.

Happily, most wine tasters regularly experience what evolving scientific understanding also proves: the importance of smell and its impact on everything from learning and loving to aging and health.

How does our sense of smell work? Why is it so evocative of emotion and so critical to enjoyment?

Our love of wine and fascination with olfaction brought us together to write this chapter. We hope you enjoy it with a good glass of wine, great memories in mind, and more in the making.

WENDY DUBIT, founder of The Senses Bureau (www.thesensesbureau.com) and Vergant Media (www.vergant.com), served as editor-in-chief of *Friends of Wine* and *Wine Enthusiast* magazines before starting a number of media entities ranging from record labels to television series. She continues to write and speak on wine, food, and lifestyle and to lead tastings worldwide, including Wine Workout, which uses tasting wine as a way to strengthen the senses, memory, and mind.

A LETTER FROM KEVIN

I have always been fascinated by the sense of smell. My earliest memories of my grandfather's farm are dominated by the sweet aroma of chamomile tea brewing on the kitchen woodstove. I grew up in a small town not far from New York City in a house surrounded by a forest of clean, fragrant pines. I am convinced their memory led to my building a new house smack in the middle of a pine forest.

I practically lived at the local swimming pool when I was a child, and smelled like chlorine all summer long. Even as an adult, I find the taste and smell of hot buttered popcorn contributes enormously to how much I enjoy a movie. Of course not everything smells like roses or summons sweet memories. We once had a gingko tree planted in our front yard that shed foul, nasty-smelling fruit once a year.

I will never forget the smell of downtown New York following the

September 11th terrorist attacks. It lingered for months. I recall an instant a year later, a sudden whiff would catch me by surprise. Seven years later, its mephitic cloud still hovers; a smell indelibly etched in my memory.

I was seduced by the smells of wine—grapes, earth, the wine itself—as a young adult; thirty years later I am still in their thrall. The sexy musk of freshly plowed earth, the fragrant balm of fermenting grapes right after harvest, and the lusty bouquet of new Beaujolais still intoxicate. An old red wine with hints of tobacco, mushrooms, and fallen leaves and the damp earth smell of my fifty-five-degree wine cellar still invigorate.

Just as the smell of cooking garlic whets my appetite and the smell of the sea shore calms my nerves, the smell of crisp autumn air reminds me that the harvest is in and my favorite time of year has come to its end.

When I was a teenager with skyrocketing hormones, I doused myself in English Leather cologne and, to this day, opening a bottle immediately transports me back to the fun-filled days of youth. I'm not sure why, but I stopped wearing cologne when I began to truly smell wine. Is it coincidence or is it smell?

Smell is critical to the preservation of life and one of our most primitive senses. Yet we humans, having radically reshaped our relationship with the natural world, often take our sense of smell for granted. I continue to examine how smell and taste figure in my life. I hope that you will enjoy exploring the fascinating mysteries of olfaction in this chapter and that it will enrich your daily life and intensify your enjoyment of wine.

A LETTER FROM WENDY

I've always lived fully attuned to each of the five senses, but it's my nose that I've followed most. Doing so has brought me nearly everywhere I want to be—most notably, to the wine industry.

Implicitly, from as far back as age eight, I linked my sense of smell to my love of learning. If my mind tired during studies, I might peel a tangerine, sit nearer a lilac bush or linden tree. By high school, I used this technique more explicitly, and began blending in color and music for optimal learning and sensory experiences. Concepts and their applications became linked to and triggered by shades and scents of citrus and pine, strands of Vivaldi and Chopin. What I lacked in photographic memory, I could usually compensate for by engaging and summoning my senses.

As a teen, I was invited to the "openings" of my parent's dinner parties, where my dad, who delighted in all things Burgundy and Bordeaux, would offer me a small pour of wine and ask for my olfactory impressions. I'll always remember the evening I discovered a marriage of river rocks, wet leather from the underside of a saddle, hay flecked with wildflowers, and golden apples in a single pour. "Ah," Dad said, nodding in encouragement, "Puligny-Montrachet." And I laughed because the name itself was as delightfully poetic to my ear as the wine had been to my nose and mouth. Even though those youthful days were rich with honest smell—kitchen, backyard, woods, stream, and farm— I'll always remember being most awed by the abundant and complex smells I found in a single glass of wine. A glass of great wine encompasses an entire world of smell, layer upon layer of riches.

Since those early dinner parties, Puligny has become a favorite wine and a symbol of the enduring bond between Dad and me. In fact, the last time I saw Dad was in a hospital—decades after those dinner parties—and I'll always cherish the evening he asked for a glass of Puligny. I was sure he knew that the orderlies wouldn't bring a bottle with his dinner, so I talked to him about the wine. I described its beguiling white Burgundian blend of minerals, acidity, crisp fruit, and toasted oak. I recalled our trips to the Côte de Nuits and the Côte de Beaune, and detailed the particular pleasures of each bottle of wine we'd brought home. As I was talking, my father became increasingly relaxed. By the time I finished he was utterly calm, with a dramatically improved oxygenation level. "Thank you," he said, bowing his head, "I just wanted to hear it in your words."

Wine is both training ground and playground for all five of our senses, our memory, and our intellect. Consequently, our enjoyment and understanding of the full wine-tasting experience improves dramatically if we observe, smell, taste, feel, analyze, and remember more carefully. Good wine demands that we stop and savor each taste to allow its rich, sensuous story to more completely unfold. Every wine deserves the attention of all five senses. But smell, with its ability to evoke memory and feeling, matters most.

SO, HOW DO WE SMELL?

With each inhalation, we gather essential information about the world around us—its delights, opportunities, and dangers. We can shut our eyes, close our mouths, withdraw our touch, and cover our ears, but the nose, with notable exceptions, is always working, alerting us to potential danger and possible pleasure.

AS PROOF OF the evolutionary importance of smell, 1 to 2 percent of our genes are involved in olfaction, approximately the same percent that is involved in the immune system.

Our sense of smell also enhances learning, evokes memory, promotes healing, cements desire, and inspires us to action. It is so important to the preservation and sustenance of life that the instantaneous information it gathers bypasses the thalamus, where other senses are processed, and moves directly to the limbic system. The limbic system controls our emotions, our emotional responses, mood, motivation, and our pain and pleasure sensations, and it is where we analyze olfactory stimuli.

Memory stored in the limbic system uniquely links emotional state with physical sensation, creating our most important and primitive form of learning: working memory. We remember smell differently than we recall sight, sound, taste, or touch because we may respond to smell the same way we respond to emotion: an increased heart rate, enhanced sensitivity, and faster breathing. It is this emotional connection that gives smell the power to stimulate memory so strongly and why a single smell can instantly transport us back to a particular time and place.

In 2004, the Nobel Prize in Medicine was awarded to Columbia University Professor Richard Axel and Hutchinson Cancer Research Center Professor Linda B. Buck for their breakthrough discoveries in olfaction. Axel and Buck discovered a large family of genes in the cells of the epithelium, or lining, of the upper part of the nose that control production of unique protein

IMPLICIT MEMORIES are perceptual, emotional, sensory, and are often unconsciously encoded and retrieved. Explicit memories are factual, episodic, temporal, and require conscious coding and retrieval. A good wine, well perceived and described, lives on in both forms of memory.

ALLERGIES, INJURY, illness, and sexual activity are just some of the reasons our noses can become temporarily or permanently clogged or occluded.

HOW DO WE SMELL?

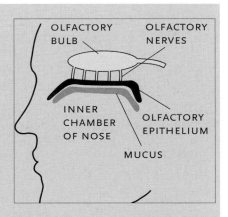

OLFACTORY BULB
OLFACTORY NERVES
INNER CHAMBER OF NOSE
OLFACTORY EPITHELIUM
MUCUS

Chemical components—esthers, ethers, aldehydes, etc.—of inhaled Puligny swirl upward through the nostrils on currents of air

Midway up the nose, millions of olfactory receptor neurons (olfactory epithelium), with their specialized protein receptors, bind the odorants that fit their specific profile

Interaction of the specific odor molecules matched with the right receptor match causes the receptor to change its shape. This change gives rise to an electrical signal that goes first to the olfactory bulbs and then to the areas of the brain that convert the electrical signal to the identification of a smell, or group of smells. The brain associates the smell(s) with perception, impressions, emotions, memories, knowledge, and more.

WHAT ARE OUR wine senses? Hearing (as in corks popping, wine pouring), seeing, smelling, tasting, feeling, and reflecting to be sure—but also more. Scientists and experts agree that smell accounts for up to 90% of what many perceive as taste and mouthfeel.

receptors, called olfactory receptors. Olfactory receptors specialize in recognizing, then attaching themselves to, thousands of specific molecules of incoming odorants. Once attached, the trapped chemical molecules are converted to electrical signals. These signals are relayed to neurons in the olfactory bulbs (there is one in each nasal cavity) before being carried along the olfactory nerve to the primary olfactory cortex, part of the brain's limbic system, for analysis and response. By the time the electrical signals of smell are directed to the limbic system the component parts of a smell—wet leather, wildflowers, golden apples, and river rocks—have already been identified and translated into electric signal. The limbic system recombines these components for analysis by scanning its vast memory data bank for related matches. Once analysis is completed, the limbic system triggers an appropriate physiological response. In the case of our Puligny wine, the limbic system might recognize it as a pleasant white wine made from Chardonnay grapes. More experienced wine tasters, with a more highly developed memory data bank, connect our wine to other Puligny wines and will recognize it as Puligny. Expert tasters might be able to recall the vineyard, maker, and year. The more we taste, test, and study, the better we become at identification.

The limbic system is the collective name for structures in the human brain involved in emotion, motivation, and emotional association with memory. It plays its role in the formation of memory by integrating emotional states with stored memories of physical sensations—such as smells.

SIZE AND SHAPE do matter. Deep, good wine glasses, such as Riedel's lines of stemware, do much to enhance varietal aroma.

THE NOSE KNOWS: Exposure to certain easily recognizable smells while forming memories will link the memory to the smell. So (as Wendy implicitly understood as a young girl) eating tangerines, for example, while studying for a test makes it easier to remember the information by recalling the smell of tangerines while taking the test.

OLFACTORY PATHWAY OF PULIGNY-MONTRACHET FROM BOTTLE TO BRAIN

We can trace the olfactory pathway of Puligny from bottle to brain, through the following steps:

- We open the bottle, in happy anticipation.
- We pour the wine into a proper glass.
- We swirl the glass to release the wine's aromas.
- We inhale the wine's bouquet deeply and repeatedly.
- Chemical components—esters, ethers, aldehydes, etc.—of the Puligny swirl upward through the nostrils on currents of air.
- Midway up the nose, millions of olfactory receptor neurons (olfactory epithelium), with their specialized protein receptors, bind the odorants that form the components of the specific wine profile.
- Interaction of the specific odor molecules matched with the right receptor causes the receptor to change shape.
- This change gives rise to an electrical signal that goes first to the olfactory bulbs and then to the areas of the brain that convert the electrical signal to the identification of a smell, or group of smells.
- The brain associates the smell(s) with perception, impressions, emotions, memories, knowledge, and more.

HOW DO WE TASTE?

Like smell, taste belongs to our chemical sensing system. Taste is detected by special structures called taste buds, and we have, on average, between five thousand and ten thousand of them, mainly on the tongue but with a few at the back of the throat and on the palate. Taste buds are the only sensory cells that are regularly replaced throughout a person's lifetime, with total regeneration taking place approximately every ten days. Scientists are examining this phenomenon, hoping that they will discover ways to replicate the process, inducing regeneration in damaged sensory and nerve cells.

Clustered within each taste bud are gustatory cells that have small gustatory hairs containing gustatory receptors. The gustatory receptors, like the olfactory receptors, are sensitive to specific types of dissolved chemicals. Everything we eat and drink must be dissolved—usually by the saliva—in order for

OLFACTORY BULB

LIMBIC SYSTEM

FIRST IMPRESSIONS? A well-crafted wine's aroma evolves in the glass. And, our noses quickly become inured to smell. This is why it's advisable to revisit wine's aroma a few times in any given tasting or flight. (Some tasters disagree with this, insisting that only the first impression matters.) Tasters can take a cue from the old perfumers' trick of sniffing their sleeves between the many essences/elixirs they may smell on a given day. In other words, they turn to something completely different—balancing sense with non-sense.

the gustatory receptors to identify its taste. Once dissolved, the gustatory receptors read then translate a food's chemical structure before converting that information to electrical signals. These electrical signals are transmitted, via the facial and glossopharyngeal nerves, through the nose and on to the brain where they are decoded and identified as a specific taste.

OUR SALIVATION

Saliva is critical not only to the digestion of food and to the maintenance of oral hygiene, but also to flavor. Saliva dissolves taste stimuli, allowing their chemistry to reach the gustatory receptor cells.

Remember being told to chew your food slowly so that you would enjoy your meal more? It's true. Taking more time to chew food and savor beverages allows more of their chemical components to dissolve and more aromas to be released. This provides more material for the gustatory and olfactory receptors to analyze, sending more complex data to the brain, which enhances perception. Taste and smell intensify.

While the majority of our taste buds are located in the mouth, we also have thousands of additional nerve endings—especially on the moist epithelial surfaces of the mouth, throat, nose, and eyes—that perceive texture, temperature, and assess a variety of factors, which recognize sensations like the prickle of sulfur, the coolness of mint, and the burn of pepper. While humans can detect an estimated ten thousand smells and smell combinations, we can taste just four or five basic flavors—sweet, salty, sour, bitter, and umami (savory). Of these, only sweet, sour, and occasional bitterness are applicable to wine tasting.

The overall word for what we perceive in food and drink through a combination of smelling, tasting, and feeling is flavor, with smell being so predominant of the three that Kevin believes that wine tasting is actually "wine smelling," and some chemists describe wine as "a tasteless liquid that is deeply fragrant." It is flavor that lets us know whether we are eating an apple or a pear, drinking a Puligny-Montrachet or an Australian Chardonnay. Anyone doubting the importance of smell in determining taste is encouraged to hold his or her nose while eating chocolate or cheese, either of which will tend to taste like chalk.

TASTE PLACE

Though maps, such as the one shown on page 260, typically place concentrated sweet receptors at the tip of the tongue, bitter at the back, and sour at the sides,

TASTING AND chewing increase the rate of salivary flow.

BITTERNESS IN wine arises from a combination of high alcohol and high tannin.

MUCH OF WHAT is commonly described as taste—80–90% or more—is aroma/bouquet as sensed and articulated by our olfactory receptors, and mouthfeel and texture as sensed by surrounding organs.

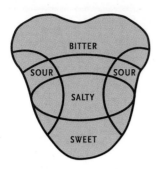

taste buds are broadly distributed throughout the mouth. In wine tasting, it is important to let wine aerate once it reaches your mouth so that it will release more aromas and intensify flavor. This is done by rolling the wine over your tongue and allowing it to linger on the tongue. This process also spreads the wine across a wider array of gustatory receptors, and gives them more time to analyze its taste. You'll discover that a good wine will reveal a first, middle, and lasting impression, which are largely determined by aroma and mouthfeel.

Mouthfeel

Mouthfeel is literally how a wine feels in the mouth. These feelings are characterized by sensations that delight, prick, and/or pain our tongue, lips, and cheeks, and that often linger in the mouth after swallowing or spitting. They can range from the piquant tingle of Champagne bubbles to the teeth-tightening astringency of tannin; from the cool expansiveness of menthol/eucalyptus to the heat of a high-alcohol red; and from the cloying sweetness of a low-acid white to the velvet coating of a rich Rhône. The physical feel of wine is also important to mouthfeel, and includes: body, thin to full; weight, light to heavy; and texture, austere, unctuous, silky, and chewy. Each contributes to wine's overall balance. More than just impressions, these qualities can trigger a physical response—drying, puckering, and salivation—which can literally have wines dancing on the tongue and clinging to the teeth.

SMELL AND TASTE TOGETHER

Recent research presents additional proof that taste seldom works alone (something wine tasters have known for centuries) and provides the first clear scientific evidence that olfaction is uniquely "dual." We often smell by inhaling through both nose and mouth simultaneously, adding to smell's complexity.

There are two paths by which smells can reach the olfactory receptors:

ORTHONASAL STIMULATION: Odor compounds (smells) reach the olfactory bulb via the "external nares" or nostrils.

RETRONASAL STIMULATION: Odor compounds reach the olfactory bulb via the "internal nares," located inside the mouth (the respiratory tract at the back of the throat). This is why even if you pinch your nose shut, a strong cheese inhaled through the nose and mouth may still smell. Molecules that stimulate the olfactory receptors float around in your mouth, up through your internal nares, and stimulate the olfactory neurons in the olfactory bulb.

According to an article in a recent issue of the journal *Neuron*, researchers reported that the smell of chocolate stimulated different brain regions when introduced into the olfactory system through the nose (orthonasally) than it did when introduced through the mouth (retronasally). The study suggests that sensing odor through the nose may help indicate the availability of food while identification through the mouth may signify receipt of food.

OLFACTORY ABILITIES OVER A LIFETIME

Are we born with equal olfactory ability? What accounts for the rise and fall of olfactory abilities over the course of a lifetime, or even over the course of a day?

Experts agree that, except in rare instances of brain disease or damage, we are born with relatively equal ability to perceive smell, although our ability to identify and articulate those smells varies widely. In general, women have the advantage over men in perceiving and identifying smells throughout their life cycle. While it's not known why, it is theorized that because women bear the enormous responsibility of conceiving and raising children, a keen sense of smell evolved as an aid to everything from choosing a mate to caring for a family.

WHO ARE YOU?

According to Janet Zimmerman, writing in Science of the Kitchen: Taste and Texture, *approximately one quarter of the population are "non-tasters," one quarter are "super-tasters," and the remaining half are "tasters." Super-tasters have a significantly higher number of taste buds than tasters, and both groups outnumber non-tasters for taste buds. The averages for the three groups are 96 taste buds per square centimeter for non-tasters, 184 for tasters, and a whopping 425 for super-tasters. Super-tasters tend to taste everything more intensely. Sweets are sweeter, bitters are bitterer, and many foods and beverages, including alcohol, taste and feel unpleasantly strong. Non-tasters are far from picky, and seem less conscious of and therefore less engaged with what they eat and drink. Tasters, the largest and least homogeneous group, vary in their personal preferences but tend to enjoy the widest array of food and drink, and to relish the act of eating and drinking the most of the three groups.*

DAYTIME DRINKING

No scientific evidence shows that our olfactory abilities change over the course of a day, although many winemakers and wine professionals believe their senses to be keener and their palates cleaner in the morning. When evaluating wines for his course and books, Kevin likes 11 A.M. For her work, Wendy favors tasting with a slight edge of hunger, which seems to enhance her alertness. Get to know your own cycles!

SUPER-TASTERS have more than 10,000 taste buds.

SENSORY OVERLOAD

Super-tasters can be supersensitive, and they may find wines with tannin and high alcohol too bitter; so Cabernet Sauvignon, for example, may not be to their liking. They may also be put off by any sweetness in wine. Non-tasters are the opposite; they might not be bothered by tannin or high alcohol, and sweet wines would probably be more acceptable.

35% OF WOMEN and 15% of men are super-tasters.

THE SMELL & TASTE TREATMENT and Research Foundation published the following lists of the different smells that arouse women and men:

Women	Men
Turn-ons	*Turn-ons*
Good & Plenty candy	Lavender
Cucumber	Pumpkin pie
Baby powder	Licorice
Banana-nut bread	Doughnuts
Lavender	Cinnamon buns
Pumpkin pie	
Chocolate	
Turn-offs	*Turn-offs*
Cherry	None found
Charcoal barbecue smoke	
Men's cologne	

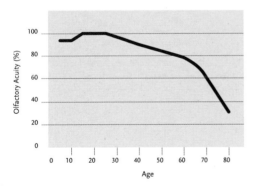

OLFACTORY ACUITY is at its peak in young adulthood.

Dr. Alan Hirsch of the Smell & Taste Treatment and Research Foundation and experts from the Monell Chemical Senses Center have shed additional light on the evolution and devolution of smell over a lifetime by describing the changes that occur at periods in the life cycle.

Fetus: In utero, the fetus acquires its blood supply through the placenta, via which it may receive odorants that help the baby recognize Mom at birth, and which may lead to developing food preferences for what Mom ate or drank.

Lactation: Flavors and smells from food and beverage make their way through mother's milk to the infant. As a result, preferences are formed toward sweet and away from bitter. Any wine and spirits the mother consumes while nursing are measurable in the baby.

Early Childhood: As a child grows, so does his or her ability to recognize and remember different odors, especially those that are paired with an emotional event. At this point in their development, children usually have a hard time describing smells in words, but they are forming lifelong positive and negative sensory and emotional impressions. For example, smelling roses in the garden with Mom will have a far different impact on the feeling the scent elicits later in life than if a child first smells roses at the funeral of a loved one.

Elementary School Years: Boys' and girls' ability to perceive and identify smells continues to develop, becoming more acute, with the ability to distinguish more smells.

Puberty: The sense of smell is at its most acute in both men and women, although women surge further ahead at the onset of menstruation. This heightened sensitivity to smell will persist throughout their fertile years.

Adulthood: Women consistently outscore men in their ability to put names to smells in adulthood, and women give higher ratings on pleasure and intensity, lower ratings on unpleasant aromas. Women's sense of smell is particularly acute at ovulation and during pregnancy.

Midlife: Men and women slowly begin to lose their acuity of smell between the ages of 35 and 40, though the ability to identify and remember smells can continue to improve over the course of a lifetime.

Age 65: By the time they reach age 65, about half the population will experience a decline of, on average, 33 percent in their olfactory abilities. A quarter of the population has no ability to smell after 65.

Age 80: A majority of the population will show losses of up to 50 percent in olfactory abilities by age 80.

ACCOUNTING FOR SMELL AND TASTE

What accounts for ethnic and cultural differences in smell and taste abilities, perceptions, and preferences? According to Monell Chemical Senses Center anthropologist Claudia Damhuis, while some odor associations/preferences might be innate (e.g., an aversion to the odor of rotten food), the vast majority of them are learned.

THE WAY WE LEARN

The ability to identify and perceive odors, as well as odor preferences, are learned both implicitly and explicitly through a number of variables including environment, culture, custom, context, and a variety of sociocultural factors.

Among the factors that impact cultural differences in odor preferences are:
- Physiological differences such as the size and presence of glands in different populations
- Differences in hygienic habits
- Acceptability and denial of odors according to cultural norms and etiquette
- Familiarity with specific odors
- The role and function attributed to specific odors
- Cultural differences in food preparations and related olfactory experiences and association
- Precision of language used to express odor perceptions and associations

IMPLICIT LEARNING is largely unconscious. Odors are identified and perceived and associations are made mainly in the background (such as the smell of hot dogs in a baseball stadium). Explicit learning is a conscious process, in which the person is aware of the learning that is taking place (such as at a wine tasting).

IN THE EMPEROR OF SCENT, Chandler Burr uses a particularly pungent Burgundian cheese to illustrate cultural preferences: When they smell the Soumaintrain, "Americans think, 'Good God!' The Japanese think, 'I must now commit suicide.' The French think, 'Where's the bread?'"

ENGLISH WINE writers will occasionally use the word gooseberries to describe a wine's aroma, but most Americans are unfamiliar with gooseberries and don't know what they smell like.

ACQUIRED TASTES

Many cultures develop their own special foods and beverages. Some, such as chili peppers, can become highly popular and spread quickly to other groups. Others are not so appealing and remain the provender of their own cultures. Think of Korea's kim chee, Norway's lutefisk, prairie oysters in the American West, and, in the wine world, Greece's retsina, a wine flavored by tree resin.

THE WAY WE SMELL

According to Lyall Watson's *Jacobson's Organ and the Remarkable Nature of Smell,* our two square meters of skin come equipped with three million sweat glands. But whereas people of European and African descent have armpits densely packed with apocrine glands, people of Asian origin have far fewer, and some have no armpit glands at all; 90 percent of the population in Japan has no detectable underarm odor. Japanese people who first encountered European traders in the nineteenth century described them as *bata-kusai*—"stinks of butter." Meantime, the French were reveling in their own aromas. Napoléon sent a note to Josephine that read: "I'll be arriving in Paris tomorrow evening. Don't wash."

GENERAL PREFERENCES

Across a great variety of ethnicities, odors can be classified into a few general categories of association, which are smells and flavors associated with nature, man, civilization, food and drink, and more.

A Worldwide Smell Survey conducted by *National Geographic* and Monell found universal unpleasantness noted across nine regions for fecal odors, human odors, material in decomposition, and mercaptans (sulfur-containing compound often added to natural gas as a warning agent). Most odors, however, were rated pleasant, including vegetation, lavender, amyl acetate (banana odor), galaxolide (synthetic musk), and eugenol (synthetic clove oil).

CULTURAL DIFFERENCES

Physiology, environment, diet, and language all play roles in cultural preference. The Asian reliance on soy rather than dairy products—which carries through even in the smells and flavors of mothers' milk—might account for Asian cultural aversion to strong cheese smells and flavors. Asian-American communities, families, or individuals may or may not exhibit this aversion, depending on the degree to which they retain native food-and-drink customs or adopt local ones.

Other cultural differences might in fact be environmental, as was suggested by a study of the Serer N'Dut of Senegal, who classify their odors in only five categories—scented, milk-fishy, rotten, urinelike, and acidic—or linguistic, as in a cross-cultural French / Vietnamese / American study in which participants' odor categorization did not correlate with their linguistic categorization.

THINGS THAT CAN AFFECT YOUR SENSE OF SMELL

According to University of California at San Diego (UCSD) Nasal Dysfunction Clinic, an estimated 1 to 2 percent of the American population suffers from the total loss of the sense of smell.

The primary complaint for the majority of people who seek help for smell loss is that food has lost its flavor. But their lives suffer in other ways, too—from the inability to avoid such dangers as spoiled food and gas leaks to

MERCAPTANS, WHICH have been described as smelling uncomfortably skunklike, are a natural by-product of fermentation. They can be removed by aeration during the winemaking process.

NUMEROUS STUDIES have proven the power of scents to affect mood and memory. Lavender has the power to calm. Citrus enhances alertness, and is now being broadcast in many office buildings in Japan. And as Shakespeare writes in *Hamlet*, "There's rosemary, that's for remembrance. Pray thee, love, remember."

HYPERGEUSIA, HYPOGEUSIA, and aguesia—heightened taste and partial and total taste loss respectively—are rare, and usually reflect a smell loss, which is often confused with a taste loss.

missing out on some of life's greatest and most sensuous pleasures: The sense of smell plays a key role in sexual excitement. And while Americans spend an enormous amount of time and money masking body odors and sexual smells, those smells remain vital to our sense of self and our bonds with others—from the families we are born into to the families we create.

WHAT CAUSES SMELL AND TASTE DISORDERS?

It is extremely rare to be born with chemosensory disorders. Most smell disorders develop after an injury or illness or in response to irritants or drugs.

An array of injuries, illnesses, and irritants can adversely affect the sense of smell, partially or fully, permanently or temporarily, including head or brain injuries; nose injuries; allergies, colds, and sinusitis; diseases including diabetes, epilepsy, and lupus; and pollutants and toxic chemicals.

Unfortunately, even when an illness is temporary or a chemical irritant is fleeting, damage to nasal plates and passages can be permanent, and can impair or destroy the sense of smell. While cells in the olfactory epithelium do regenerate normally, severe damage can eliminate their ability to regenerate.

Treatments that can adversely affect the sense of smell include surgery of the head and neck; dental work; chemotherapy and other cancer treatments; many types of medications; and alcohol.

Some medicines can have a positive impact on the sense of smell and taste. Interestingly, groups of medicines used to treat allergies, sinusitis, infection, and inflammation can open nasal passages and therefore have temporary ameliorative effects on smell and taste. They include antihistamines and decongestants, antibiotics, and nasal douches and lubricants.

Cluster headaches, migraines, and Addison's disease can temporarily heighten (sometimes horrifically) the sense of smell and taste.

SMELL AND TASTE TESTING AND TREATMENT

There are many interesting ways for professionals and lay people alike to test the sense of smell. Richard Doty, Ph.D., founder of Sensonics, Inc., has developed an array of tests that assess chemosensory function. They range from the scratch-and-sniff University of Pennsylvania Smell Identification Test to the Smell Threshold Test, which is often used to identify sensory experts and to screen workers involved with potential hazards. The Smell and Taste Treatment and Research Foundation uses these, as well as Hedonic Testing, to

I CAN'T GET NO OLFACTION

Osme, the Greek word meaning "odor," is the root of many medical definitions relating to smell, among them:

NORMOSMIA: Normal sense of smell

DYSOSMIA: Any defect or impairment in the sense of smell; a collective term that includes any of the following:

ANOSMIA: Complete loss of the sense of smell

HYPOSMIA: Partial loss of the sense of smell

HYPEROSMIA: Increased ability to smell

PAROSMIA: Distortion of the sense of smell

PHANTOSMIA: Phantom smells and olfactory hallucinations

PRESBYOSMIA: A decrease in the sense of smell associated with aging

A DEVIATED SEPTUM, which can cause problems with proper breathing and nasal discharge, is an abnormal configuration of the cartilage that divides the two sides of the nasal cavity. While definitive treatment may require surgery, it turns out that decongestants, antihistamines, nasal cortisone spray, nasal lavage, and even eating jalapeño peppers and wasabi (which are hot enough to flush a stopped-up nose) can be temporarily helpful.

SMELLING TOO WELL?

Alan Hirsch of the Smell & Taste Treatment and Research Foundation tells of a woman with Addison's disease whose sense of smell was heightened 1000 percent—approaching the olfactory acuity of a bloodhound or cockroach—by the illness. The smells she picked up bothered her immensely. The patient developed agoraphobia (a fear of being in crowds or public places) and was not able to leave her home. Her heightened abilities (or in this case, disabilities) came under control when the disease did.

Ben Cohen of Ben & Jerry's is purportedly hyposmic. As often happens with sensory loss, other senses compensate where possible. Might this be why he created Chunky Monkey and so many richly textured ice creams in the Ben & Jerry's line?

SOME YOGA traditions use a neti pot, a kind of nasal douche, for keeping the nose's mucous layer moist and healthy.

ANTIHISTAMINES, while they may clear the nasal passages, also tend to dry them out, so they can be a mixed blessing.

LANGUAGE OF SMELL

Finding the language to describe what we smell, and how what we smell affects us, evolves over our lifetime, with women being slightly better at it than men.

measure how much subjects like what they smell, and Memory Testing to see what types of memory (short-term, long-term, implicit, explicit) the smells have been committed to. More information on these and other methods of assessing the olfactory sense can be found by consulting the resources at the end of this chapter.

Because smell can be so important to one's lifestyle, it is not uncommon to find coexisting mood disorders among people with olfactory disorders. Depending on the cause and severity of the smell disorder, treatment ranges from medications such as vitamins, steroids, calcium channel blockers, anti-convulsants, and antidepressants, to surgery. It is important to seek help to identify, assess, treat, and manage these disorders.

For all those who can, we recommend . . . wine tasting!

TRAINING AND STRENGTHENING

Many animals may have a more highly developed sense of smell than we do. But we humans have brains brilliantly designed to preserve, protect, and enrich all aspects of our lives through smell. In our experience, there is no better way to train and strengthen the senses, memory, and mind than through fully, consciously savoring wine.

The best news? Besides investing in some good bottles, proper storage, and proper glassware, all we need to fully appreciate wine is built in, in the form of our five senses. Wine tasters have known innately for ages what evolving scientific understanding and empiric evidence are only now proving: the power and importance of smell. And while the explicit knowledge of the science of smell is good, it is not always essential to our implicit experience.

That said, to embrace, apply, articulate, and improve on what we already have, let's open a bottle of vino!

TOWARD A COMMON LANGUAGE OF TASTE AND SMELL

Smell is a relatively inadequate word for our most primitive and powerful sense. It means both the smells that emanate from us (we are what we eat and drink) as well as the smells we perceive. Throughout history, wine tasters have done much to create a common language, and to savor the intersection where enlivened and articulated senses meet memory, anticipation, association, and personal preferences throughout history.

Like colors, aroma can be broken down into basic categories which, when combined, yield the rich symphony that is wine. The UC Davis's Wine Aroma Wheel (see Resources for more information) categorizes basic fruit aromas as citrus (grapefruit, lemon), berry (blackberry, raspberry, strawberry, black currant), tree (cherry, apricot, peach, apple), tropical (pineapple, melon, banana), dried (raisin, prune, fig), and others. Likewise, vegetative aromas can be categorized as fresh (stemmy, grassy, green, eucalyptus, mint), canned (asparagus, olive, artichoke), and dried (hay, straw, tea, tobacco). Other aroma categories include nutty, caramelized, woody, earthy, chemical, pungent, floral, and spicy.

Still, no two people are alike in either how they smell or the smells they perceive. It is deeply personal and experiential. Here is our best effort to convey how scent feels and what it means to us.

Wendy says *Puligny* to herself, simply because she loves the word nearly as much as the wine. It serves as homing signal that brings her to self and center—to a time and a place and a wine she loves, and to the kind of bonds that endure.

Kevin goes out on his porch with a glass of some favorite, special vintage and looks into the night sky. Embodied in every sip is all that he treasures and honors about Windows on the World, and a reason to look up at stars.

We're hoping and trusting that you'll find wine and spirits like this, and that you'll share words of them with us. Meantime, here's to your health and happiness, and to savoring wine and life in and with every sense!

DIFFERENT SMELLS OF WINE COME FROM:
The grape
The winemaking
The ageing

SOURCES AND RESOURCES

Monell Chemical Senses Center, www.monell.org

National Institute of Health, www.nidcd.nih.gov/health/smelltaste

Professional Friends of Wine: A Sensory User's Manual, www.winepros.org/wine101/sensory_guide.htm

Sensonics, www.sensonics.com

Sense of Smell Institute, www.senseofsmellinstitute.org

Smell & Taste Treatment and Research Foundation, www.scienceofsmell.com

The Senses Bureau, www.thesensesbureau.com

Tim Jacob Smell Research Laboratory, www.cf.ac.uk/biosi/staff/jacob

UCSD Nasal Dysfunction Clinic, www.surgery.ucsd.edu/ent/DAVIDSON/NDC/booklet.htm

University of California Davis Wine Aroma Wheel, www.winearomawheel.com

A Natural History of the Senses by Diane Ackerman

Jacobson's Organ and the Remarkable Nature of Smell by Lyall Watson

Life's A Smelling Success: Using Scent to Empower Your Memory and Learning by Alan Hirsch, M.D.

The Emperor of Scent: A Story of Perfume, Obsession, and the Last Mystery of the Senses by Chandler Burr

Matching Wine and Food

BY KEVIN ZRALY AND ANDREA IMMER

ANDREA IMMER WORKED WITH KEVIN AT WINDOWS ON THE WORLD. SHE is one of only 14 women in the world to hold the title of Master Sommelier, awarded by the Court of Master Sommeliers. She has written several books on wine, and was named Outstanding Wine & Spirits Professional by the James Beard Foundation.

YOU'VE JUST TASTED your way through eight classes in this book and discovered at least a shopping cart's worth of wines to really enjoy. And for what purpose? Food! The final stop on the wine odyssey—and the whole point of the trip—is the dinner table. Quite simply, wine and food were meant for each other. Just look at the dining habits of the world's best eaters (the French, the Italians, the Spanish): Wine is the seasoning that livens up even everyday dishes. Salt and pepper shakers are a fixture of the American table, but in Europe it's the wine bottle.

WINE-AND-FOOD MATCHING BASICS

ARE YOU a menu maven or a wine-list junkie? Personally, I look at the wine list first, choose my wine, and then make my meal selection.

First, forget everything you've ever heard about wine-and-food pairing. There's only one rule when it comes to matching wine and food: The best wine to pair with your meal is whatever wine you like. No matter what!

If you know what you want, by all means have it. Worried that your preference of a Chardonnay with a sirloin steak might not seem "right"? Remember, it's your own palate that you have to please.

CHARDONNAY IS a red wine masquerading as a white wine, which, in my opinion, makes it a perfect match for steak.

What's wine-and-food synergy?

Sounds like a computer game for gourmets, right? If up until now you haven't been the wine-with-dinner type, you're in for a great adventure. Remember, the European tradition of wine with meals was not the result of a shortage of milk or iced tea. Rather, it results from what I call wine-and-food synergy—when the two are paired, both taste better.

HOW DO I make my wine and food decision?
 1. What kind of wine do I like?
 2. Texture of food (heavy or light)
 3. Preparation (grilled, sautéed, baked, etc.)
 4. Sauce (cream, tomato, wine, etc.)

How does it work? In the same way that combining certain foods improves their overall taste. For example, you squeeze fresh lemon onto your oysters, or grate Parmesan cheese over spaghetti marinara, because it's the combination of flavors that makes the dish.

Apply that idea to wine-and-food pairing; foods and wines have different flavors, textures, and aromas. Matching them can give you a new, more interesting flavor than you would get if you were washing down your dinner with, say, milk (unless you were dining on chocolate-chip cookies). The more flavorful the food, the more flavorful the wine should be.

Do I have to be a wine expert to choose enjoyable wine-and-food matches?

Why not just use what you already know? Most of us have been tasting and testing the flavors, aromas, and textures of foods since before we got our first teeth, so we're all food experts! As we'll show you, just some basic information about wine and food styles is all you'll need to pick wines that can enhance your meals.

What about acidity?

Acid acts as a turbocharger for flavor. A high-acid wine is a good choice for dishes with cream or cheese sauces. It enhances and lengthens the flavor of the dish. Watch television's Food Network. The TV chefs are always using lemons and limes—acidic ingredients. Even dishes that aren't "sour" have a touch of an acid ingredient to pump up the flavor. As chef Emeril Lagasse says, "Kick it up a notch!"

What role does texture play?

There's an obvious difference in the texture or firmness of different foods. Wine also has texture, and there are nuances of flavor in a wine that can make it an adequate, outstanding, or unforgettable selection with the meal. Very full-style wines have a mouth-filling texture and bold, rich flavors that make your palate sit up and take notice. But when it comes to food, these wines tend to overwhelm most delicate dishes (and clash with boldly flavored ones). Remember, we're looking for harmony and balance. A general rule is: The sturdier or fuller in flavor the food, the more full-bodied the wine should be. For foods that are milder the best wines to use would be medium- or light-bodied.

Once you get to know the wines, matching them with food is no mystery. Here is a list with some suggestions based on the texture of the wines and the foods they can match.

A FAIL-SAFE food: When in doubt, order roast chicken, which acts as a blank canvas for almost any wine style—light-, medium-, or full-bodied.

DO YOU drink your tea with milk or lemon? The milk coats your mouth with sweetness, whereas the lemon leaves your tongue with a dry crispness.

KEEP IT SIMPLE

Probably one of the reasons that classic French cuisine is noted for its subtlety is because the French want to let their wines "show off." This is an especially good idea if the wine is a special or "splurge" bottle.

White Wines

MY FAVORITE white wine for picnics is German Riesling Kabinett/Spätlese. On a hot summer day, I can think of no better white wine than a chilled German Riesling. The balance of fruit, acid, and sweetness as well as the lightness (low alcohol) make these wines a perfect match for salads, fruits, and cheese. For those who prefer a drier style Riesling, try Alsace, Washington State, or the Finger Lakes region of New York.

LIGHT-BODIED WHITES	MEDIUM-BODIED WHITES	FULL-BODIED WHITES
Alsace Pinot Blanc	Pouilly-Fumé	Chardonnay*
Alsace Riesling	Sancerre	Chablis Grand Cru
Chablis	White Graves	Meursault
Muscadet	Chablis Premier Cru	Chassagne-Montrachet
German Kabinett and Spätlese	Mâcon-Villages	Puligny-Montrachet
Sauvignon Blanc*	Pouilly-Fuissé	Viognier
Orvieto	St-Véran	
Soave	Montagny	
Verdicchio	Sauvignon Blanc*/ Fumé Blanc	
Frascati	Chardonnay*	
Pinot Grigio	Gavi	
Pinot Gris	Gewürztraminer	
	Gruner Veltliner	

Matching Foods

Sole	Snapper	Salmon
Flounder	Bass	Tuna
Clams	Shrimp	Swordfish
Oysters	Scallops	Lobster
	Veal paillard	Duck
		Roast chicken
		Sirloin steak

* Note that starred wines are listed more than once. That's because they can be vinified in a range of styles from light to full texture, depending on the producer. When buying these, if you don't know the style of a particular winery, it's a good idea to ask the restaurant server or wine merchant for help.

Red Wines

LIGHT-BODIED REDS	MEDIUM-BODIED REDS	FULL-BODIED REDS
Bardolino	Cru Beaujolais	Barbaresco
Valpolicella	Côtes du Rhône	Barolo
Chianti	Crozes-Hermitage	Bordeaux
Rioja-Crianza	Burgundy Premiers	(great châteaux)
Beaujolais	and Grands Crus	Châteauneuf-du-Pape
Beaujolais-Villages	Bordeaux (Crus Bourgeois)	
Burgundy (Village)	Cabernet Sauvignon*	Hermitage
Bordeaux (proprietary)	Merlot*	Cabernet Sauvignon*
Pinot Noir*	Zinfandel*	Merlot*
	Chianti Classico Riserva	Zinfandel*
	Dolcetto	Syrah/Shiraz*
	Barbera	Malbec*
	Rioja Reserva and	
	Gran Reserva	
	Syrah/Shiraz*	
	Pinot Noir*	
	Malbec*	

Matching Foods

Salmon	Game birds	Lamb chops
Tuna	Veal chops	Leg of lamb
Swordfish	Pork chops	Beefsteak (sirloin)
Duck		Game meats
Roast chicken		

MY FAVORITE red wine for picnics is Beaujolais. I'll never forget my first summer in France, sitting outside a bistro in Paris and being served a Beaujolais. A great Beaujolais is the essence of fresh fruit without the tannins, and its higher acidity blends nicely with all picnic fare. For barbecued shrimp in the middle of the summer, I opt for chilled Beaujolais.

MY FAVORITE red wine for lunch is Pinot Noir. Since most of us have to go back to work after lunch, the light, easy-drinking style of a Pinot Noir will not overpower the usual luncheon fare of soups, salads, and sandwiches.

MY FAVORITE wine for lamb is Bordeaux or California Cabernet Sauvignon. In Bordeaux they have lamb with breakfast, lunch, and dinner! Lamb has such a strong flavor, it needs a strong wine. The big, full-bodied Cabernet Sauvignons from California and Bordeaux blend in perfectly.

PINOT NOIR is a white wine masquerading as a red wine, which makes it a perfect wine for fish and fowl. Other choices include Chianti Classico and Spanish Riojas (Crianza and Reserva).

BITTERNESS IN wine comes from the combination of high tannin and high alcohol, and these wines are best served with food that is either grilled, charcoaled, or blackened.

COOKING WITH WINE

Try to use the same wine or style that you are going to serve.

"I cook with wine; sometimes I even add it to food."

—W. C. FIELDS

SPICY SAUCES

Try something with carbon dioxide, such as a Champagne or a sparkling wine.

THE #1 WINE-DRINKING DAY IN AMERICA

"What's your favorite wine for Thanksgiving?" is one of the most asked wine-and-food questions that I don't have a definitive answer for. The problem with Thanksgiving is that not just the turkey but everything else that is served with it (sweet potatoes, cranberry, butternut squash, stuffing) can create havoc with wine. This is also the big family holiday in the United States, and do you really want to share your best wines with Uncle Joe and Aunt Carole? Try "user-friendly" wines—easy drinking, inexpensive wines from reliable producers.

Do sauces play a major role when you're matching wine and food?

Yes, because the sauce can change or define the entire taste and texture of a dish. Is the sauce acidic? Heavy? Spicy? Subtly flavored foods let the wine play the starring role. Dishes with bold, spicy ingredients can overpower the flavor nuances and complexity that distinguish a great wine.

Let's consider the effect sauces can have on a simple boneless breast of chicken. A very simply prepared chicken paillard might match well with a light-bodied white wine. If you add a rich cream sauce or a cheese sauce, then you might prefer a high acid, medium-bodied or even a full-bodied white wine. A red tomato-based sauce, such as a marinara, might call for a light-bodied red wine.

No-Fault Wine Insurance

Drinking wine with your meals should add enjoyment, not stress, but it happens all too often. You briefly eye the wine list or scan the wine-shop shelf, thinking well, maybe . . . a beer. In the face of so many choices, you end up going with the familiar. But it can be easy to choose a wine to enjoy with your meal.

From endless experimentation at home and in the restaurant, I've come up with a list of "user-friendly" wines that will go nicely with virtually any dish. What these wines have in common is that they are light- to medium-bodied, and they have ample fruit and acidity. The idea here is that you will get a harmonious balance of flavors from both the wine and the food, with neither overwhelming the other. Also, if you want the dish to play center stage, your best bets are wines from this list.

User-Friendly Wines

ROSÉ WINES	WHITE WINES	RED WINES
Virtually any rosé or white Zinfandel	Pinot Grigio	Chianti Classico
	Sauvignon Blanc/ Fumé Blanc	Rioja Crianza
	German Riesling, Kabinett, and Spätlese	Beaujolais-Villages
	Pouilly-Fumé and Sancerre	Côtes du Rhône
	Mâcon-Villages	Pinot Noir
	Champagne and sparkling wines	Merlot

These wines work well for what I call "restaurant roulette"—where one diner orders fish, another orders meat, and so on. They can also match well with distinctively spiced ethnic foods that might otherwise clash with a full-flavored wine. And, of course, all these wines are enjoyable to drink on their own.

Wine and Cheese—Friends or Foes?

As in all matters of taste, the topic of wine and food comes with its share of controversy and debate. Where it's especially heated is on the subject of matching wine and cheese.

Wine and cheese are "naturals" for each other. For me, a good cheese and a good wine will enhance the flavors and complexities of both. Also, the protein in cheese will soften the tannins in a red wine.

The key to this match is in carefully selecting the cheese; therein lies the controversy. Some chefs and wine-and-food experts caution that some of the most popular cheeses for eating are the least appropriate for wine because they overpower it—a ripe cheese like Brie is a classic example.

The "keep-it-simple" approach applies again here. I find that the best cheeses for wines are the following: Parmigiano-Reggiano, fresh mozzarella, Pecorino, Taleggio, and Fontina from Italy; Chèvre, Montrachet, Tomme, and Gruyère from France; Dutch Gouda; English or domestic Cheddar; domestic aged or fresh goat cheese and Monterey Jack; and Manchego from Spain.

My favorite wine-and-cheese matches:

Chèvre/fresh goat cheese—Sancerre, Sauvignon Blanc

Montrachet, aged (dry) Monterey Jack—Cabernet Sauvignon, Bordeaux

Pecorino or Parmigiano-Reggiano—Chianti Classico Riserva, Brunello di Montalcino, Cabernet Sauvignon, Bordeaux, Barolo, and Amarone

Manchego—Rioja, Brunello di Montalcino

What to drink with Brie? Try Champagne or sparkling wine. And blue cheeses, because of their strong flavor, overpower most wines except—get ready for this—dessert wines! The classic (and truly delicious) matches are Roquefort cheese with French Sauternes, and Stilton cheese with Port.

MY FAVORITE cheese with wine is Parmigiano-Reggiano. Now we are really getting personal! I love Italian food, wine, and women (I even married one), but Parmigiano-Reggiano is not just to have with Italian wine. It also goes extremely well with Bordeaux, California Cabernets, and even the lighter Pinot Noirs.

SWEET SATISFACTION: WINE WITH DESSERT, WINE AS DESSERT

I remember my first taste of a dessert wine—a Sauternes from France's Bordeaux region. It was magical! Then there are also Port, Cream Sherry, Beerenauslese, to name a few—all very different wines with one thing in common: sweetness. Hence the name "dessert" wines—their sweetness closes your palate and makes you feel satisfied after a good meal. But with wines like these, dessert is just one part of the wine-and-food story.

"Wine with dessert?" you're thinking. At least in this country, coffee is more common, a glass of brandy or liqueur if you're splurging. But as more and more restaurants add dessert wines to their by-the-glass offerings, perhaps the popularity will grow for these kinds of wine. (Because they're so rich, a full bottle of dessert wine isn't practical unless several people are sharing it. For serving at home, dessert wines in half-bottles are a good alternative.)

I like a dessert wine a few minutes before the dessert itself, to prepare you for what is to come. But you certainly can serve a sweet dessert wine with the course.

Here are some of my favorite wine-and-dessert combinations:

MY FAVORITE wine for chocolate is Port. For me, both chocolate and Port mean the end of the meal. They are both rich, sweet, satisfying, and sometimes even decadent together.

Port—dark chocolate desserts, walnuts, poached pears, Stilton cheese

Madeira—milk chocolate, nut tarts, crème caramel, coffee- or mocha-flavored desserts

Pedro Ximénez Sherry—vanilla ice cream (with the wine poured over it), raisin-nut cakes, desserts containing figs or dried fruits

Beerenauslese and Late Harvest Riesling—fruit tarts, crème brûlée, almond cookies

Sauternes—fruit tarts, poached fruits, crème brûlée, caramel and hazelnut desserts, Roquefort cheese

Muscat Beaumes-de-Venise—crème brûlée, fresh fruit, fruit sorbets, lemon tart

Asti Spumante—fresh fruits, biscotti

Vouvray—fruit tarts, fresh fruits

Vin Santo—biscotti (for dipping in the wine)

Often, I prefer to serve the dessert wine as dessert. That way I can concentrate on savoring the complex and delicious flavors with a clear palate. It's especially convenient at home—all you have to do to serve your guests an exotic dessert is pull a cork! And if you're counting calories, a glass of dessert wine can give you the satisfying sweetness of dessert with a lot less bulk (and zero fat!).

How much wine should I order for my dinner party?

At a dinner party where several wines will be served, I allow one bottle for every five people, which equals approximately one five-ounce glass per person.

What's the best wine to serve with hors d'oeuvres for my dinner reception?

Champagne. One of the most versatile wines produced in the world, Champagne has a "magical effect" on guests. Whether served at a wedding or a dinner at home, Champagne remains a symbol of celebration, romance, prosperity, and fun.

FOR FURTHER READING

I recommend *Perfect Pairings* by Evan Goldstein and *Great Wine & Food Made Simple* by Andrea Immer.

12 BOTTLES OF WINE = 1 CASE

BOTTLE SIZES

375 ml = 12.7 oz = half-bottle
750 ml = 25.4 oz = full bottle
1.5 liters = 50.8 oz = magnum (two bottles)

Wine-Buying Strategies for Your Wine Cellar

Buying and selecting wines for your cellar is the most fun and interesting part of wine appreciation—besides drinking it, of course! You've done all your studying and reading on the wines you like, and now you go out to your favorite wine store to banter with the owner or wine manager. You already have an idea what you can spend and how many bottles you can safely store until they're paired with your favorite foods and friends.

Wine buying has changed dramatically over the last twenty years. Many liquor stores have become wine-specialty stores, and both the consumer and retailer are much more knowledgeable. Even twenty years ago, the wines of South Africa, Spain, New Zealand, Australia, and Chile were not the wines the consumer cared to buy. Back then, the major players were the wines of California, France, and Italy. Today there's so much more diversity in wine styles and wine prices, it's almost impossible to keep up with every new wine and new vintage that comes on the market. You can subscribe, among many publications, to *Wine Spectator*, *The Wine Enthusiast*, *Wine & Spirits*, or *Wine Advocate*, Robert M. Parker Jr.'s newsletter, to help you with your choices, but ultimately you'll find the style of wine to suit your own personal taste.

In this book I don't recommend specific wines from specific years because I don't believe that everyone will enjoy the same wines, or that everyone has the same taste buds as I do. I think it's very important that every year the consumers have a general knowledge of wineries that have consistently made great value wine, and know what's hot and what's not. Here are some of my thoughts and strategies for buying wine this year.

There is, and will continue to be, an abundance of fine wine over the next few years. The vintage years of 2000, 2003, and 2005 in Bordeaux; 2001 and 2004 in Piedmont and Tuscany; 2003, 2004, and 2005 in Germany; 2003 and 2005 in the Rhône Valley; 2002, 2003, and 2005 in Burgundy reds; 2003 in Chile and Argentina; and 2002, 2004, and 2005 in Australian Shiraz will give us great wines to drink over the next few years. The 2001, 2002, 2003, 2004, and 2005 vintages for California Cabernet Sauvignon and Chardonnay are generally excellent. The 1994, 2000, and 2003 vintage Ports are readily available. Although many of the wines from these regions are high-priced, there still remain hundreds of wines under twenty dollars that you can drink now or cellar for the future.

Anyone can buy expensive wines! In an average year I taste some three thousand wines. The real challenge is finding the best values—the ten-dollar bottle that tastes like a twenty-dollar bottle. The following is a list of my buying strategies for my own wine cellar. This is by no means a complete roster of great wines, but most are retasted every year and have been consistently good.

Everyday Wines

($10 and under)

Argentina
Alamos Chardonnay
Bodegas Norton Malbec
Bodegas Esmeralda Malbec
Trapiche Malbec
Valentin Bianchi Malbec

Australia
Alice White Cabernet Sauvignon
Jacob's Creek Shiraz/Cabernet
Lindemans Chardonnay Bin 65
Oxford Landing Sauvignon Blanc
Penfolds Sémillon/Chardonnay
Rosemount Chardonnay or
Shiraz/Cabernet (Diamond Label)
Yellow Tail

California
Beringer Merlot Founders Estate
Buena Vista Sauvignon Blanc
Fetzer Merlot Eagle Peak or
Sauvignon Blanc or Syrah
Forest Glen Cabernet Sauvignon
or Shiraz
Forest Ville Selections Cabernet
Sauvignon or Chardonnay
Glass Mountain Cabernet Sauvignon
Monterey Vineyard Cabernet Sauvignon
Napa Ridge Merlot

Pepperwood Grove Chardonnay
R.H. Phillips Cabernet Sauvignon
Barrel Cuvée
Rutherford Ranch Chardonnay

Chile
Viña Caliterra Cabernet Sauvignon
or Merlot
Carmen Carmenere
Viña Santa Rita Cabernet Sauvignon 120
Walnut Crest Merlot
Viña Montes Cabernet Sauvignon

France
Beaujolais-Villages Louis Jadot
or Georges Duboeuf
Château Bonnet Blanc
Côtes du Rhône Guigal
Côtes du Rhône Parallele "45" Jaboulet
Côtes du Rhône, Perrin
Fortant de France Merlot or Syrah
Hugel Gentil
J. Vidal-Fleury Côtes du Rhône
La Vieille Ferme Côtes du Ventoux
Les Jemelles Syrah or Merlot
Louis Latour Ardèche Chardonnay
Mâcon-Villages (most producers)
Michel Lynch
Réserve St-Martin Merlot

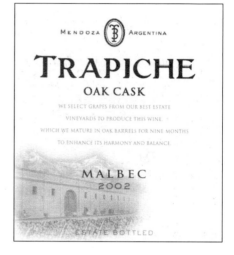

Italy
Michele Chiarlo Barbera d'Asti
Montepulciano Red, Casal Thaulero

Spain
Bodegas Montecillo Cumbrero
Marqués de Cáceres Crianza

Washington State
Columbia Crest Sémillon-
Chardonnay
Covey Run Fumé Blanc
Hogue Chardonnay Columbia Valley
Hogue Fumé Blanc

Once-a-Week Wines
($10 to $20)

Argentina
Bodegas Cantena Zapata
Malbec Alamos
Bodegas Weinert Carrascal
Navarro Correas Cabernet Sauvignon
"Collection Privada"

Australia
Banrock Station Chardonnay
Black Opal Cabernet Sauvignon
and Shiraz
Lindemans Shiraz Bin 50 or
Cabernet Sauvignon Bin 45
Penfolds Bin 389
Penfolds Chardonnay, Shiraz,
and Cabernet Sauvignon
Peter Lehmann Shiraz Barossa
Rosemount Show Reserve
Chardonnay
Wolf Blass Chardonnay

California
Amberhill Cabernet Sauvignon
Beaulieu Merlot Costal
Benziger Chardonnay, Merlot,

or Cabernet Sauvignon
Beringer Cabernet Sauvignon
Founders Estate
Beringer Chardonnay, Merlot,
and Pinot Noir
Brander Sauvignon Blanc
Calera Central Coast Chardonnay
Carmenet Cabernet Sauvignon
Castle Rock Pinot Noir
Souverain Chardonnay
and Merlot
Chateau St. Jean Chardonnay
or Sauvignon Blanc
Cline Cellars Zinfandel
Clos du Bois Chardonnay
Eshcol Cabernet and Chardonnay
Estancia Chardonnay
or Cabernet Sauvignon
Ferrari-Carano Fumé Blanc
Fetzer Valley Oaks Cabernet
Sauvignon, Zinfandel,
or Chardonnay
Forest Glen Merlot
Frog's Leap Sauvignon Blanc
Gallo of Sonoma Chardonnay,

Cabernet Sauvignon, Pinot Noir, and Merlot

Geyser Peak Sauvignon Blanc

Girard Sauvignon Blanc

Hawk Crest Chardonnay, Cabernet Sauvignon, or Merlot

Hess Select Chardonnay or Cabernet Sauvignon

Honig Sauvignon Blanc

Kendall-Jackson Chardonnay Vintners Reserve, Syrah, or Cabernet Sauvignon

Kenwood Sauvignon Blanc

Laurel Glen Quintana Cabernet Sauvignon

Liberty School Cabernet Sauvignon

Markham Merlot and Sauvignon Blanc

Mason Sauvignon Blanc

Meridian Chardonnay or Cabernet Sauvignon

Merryvale Chardonnay Starmont

Ravenswood Zinfandel (Sonoma)

Ridge Zinfandel (Sonoma)

Robert Mondavi Private Selections

Rosenblum Zinfandel Vintners Cuvée

Rutherford Vintners Cabernet Sauvignon or Merlot

Saintsbury Chardonnay

Sebastiani Chardonnay or Cabernet Sauvignon

Seghesio Sonoma Zinfandel

Silverado Sauvignon Blanc

Simi Cabernet Sauvignon or

Sauvignon Blanc or Chardonnay

St. Francis Merlot

St. Francis Old Vines Zinfandel

St. Supery Sauvignon Blanc

Chile

Casa Lapostolle Cabernet Sauvignon or Merlot Cuvée Alexandre

Concha y Toro Casillero del Diablo, Cabernet Sauvignon, Puente Alto

Cousiño Macul Antiguas-Reserva Cabernet Sauvignon

Los Vascos Reserve Cabernet Sauvignon

Viña Montes Merlot

Viña Santa Rita Cabernet Sauvignon Casa Real

France

Alsace Riesling—Trimbach, Hugel, or Zind-Humbrecht

Château de Bellevue Morgon, Jadot

Château de Sancerre, Lapostolle

Château Greysac

Château La Cardonne

Château Larose Trintaudon

Crozes-Hermitage Les Jalets, Jaboulet

Domaine Leroy Bourgogne Rouge

Drouhin Vero Chardonnay or Pinot Noir

Fleurie, Georges Deboeuf

Muscadet, Sauvion & Fils

Pouilly-Fuissé, Georges Deboeuf

St-Véran, Louis Jadot

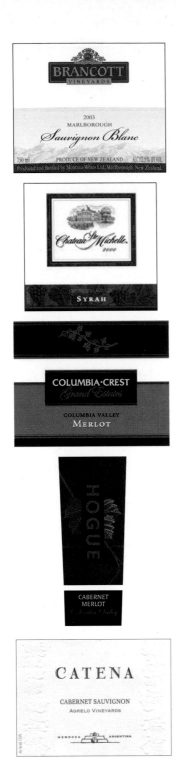

Germany
Niersteiner Olberg by Strub

Italy
Allegrini Valpolicella Classico
Anselmi Soave
Antinori Sangiovese Santa Cristina
Bollini Trentino Pinot Grigio
Boscaini Pinot Grigio
Castello Banfi Toscana Centine
Frescobaldi Chianti Ruffina
Nipozzano Reserva
Lungarotti Rubesco
Marco Felluga Pinot Grigio, Collio
Pighin Pinot Grigio
Rosso di Montalcino Col d'Orcia
Taurino Salice Salentino
Zenato Valpolicia

New Zealand
Babich Sauvignon Blanc
Brancott Chardonnay or
Sauvignon Blanc
Glazebrook Sauvignon Blanc
Goldwater Sauvignon Blanc
Kim Crawford Sauvignon Blanc
Nobilo Sauvignon Blanc
Oyster Bay Sauvignon Blanc
Saint Clair Sauvignon Blanc
Stoneleigh Chardonnay and
Sauvignon Blanc

Oregon
Argyle Chardonnay
A to Z Wineworks Pinot Noir, Pinot
Gris, Chardonnay, and Pinot Blanc
Cooper Mountain Pinot Noir
King Estate Pinot Gris
Willamette Valley Vineyards Pinot Noir

Ports
Fonseca Bin #27
Sandeman Founders Reserve

Spain
Conde de Valdemar Crianza

Sparkling Wines
Bouvet Brut
Codorniu Brut Classico
Cristalino Brut
Domaine Chandon
Freixenet Brut
Gloria Ferrer Brut
Gruet
Korbel
Scharffenberger
Segura Viudas Brut

Washington State
Columbia Crest Chardonnay, Shiraz,
Merlot, and Cabernet Sauvignon
Covey Run Chardonnay and Merlot
Hogue Cabernet Sauvignon, Pinot
Gris, and Merlot

Once-a-Month Wines

($20 to $50; most under $40)

Argentina
Catena Cabernet Sauvignon Alta
Salentein Malbec

Australia
Greg Norman Estate Shiraz
Penfolds Shiraz Bin 28

California
Cabernet Sauvignon
Artesa
Beaulieu Rutherford
Beringer Knights Valley
Clos du Val
The Hess Collection
Geyser Peak Reserve
Jordan
Joseph Phelps
Louis Martini
Mondavi
Raymond
Ridge
Turnbull
Whitehall Lane

Chardonnay
Arrowood Grand Archer
Beringer Private Reserve
Chalone
Cuvaison
Ferrari-Carano
Kendall-Jackson Grand Reserve
Mondavi
Sonoma-Cutrer

Merlot
Clos du Bois

Frei Brothers
Shafer
Phelps

Pinot Noir
Acacia
Au Bon Climat
Byron
Calera
Etude
La Crema (Anderson Valley)
Mondavi
Saintsbury (Carneros)
Truchard
Williams Selyem (Sonoma Coast)

Sparkling Wine
Chandon Reserve
Domaine Carneros
Iron Horse
Roederer Estate

Syrah
Clos du Bois
Fess Parker
Four Vines
Justin

Zinfandel
Ridge Geyserville
Rosenblum-Continente
Seghesio Old Vine

Champagne
Any non-vintage

Chile
Concha y Toro Cabernet Sauvignon
Don Melchor

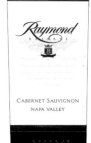

Errazúriz Don Maximiano
Founders Reserve
Veramonte Primus
Viña Montes Cabernet
Sauvignon Apalta

France
Château Carbonnieux Blanc
Château de Malle (Sauternes)
Château de Sales
Château Fourcas-Hosten
Château Fuissé (Pouilly-Fuissé)
Château Gloria
Château Lagrezette (Cahors)
Château Les Ormes de Pez
Château Meyney
Château Olivier Blanc
Château Phélan-Ségur
Château Pontesac
Château Sociando-Mallet
Crozes Hermitage Domaine
de Thalabert, Jaboulet
Ladoucette Pouilly-Fumé
Mercurey Faiveley
Olivier Leflaive Puligny-Montrachet
Pascal Jolivet (Pouilly-Fumé)

Germany
Kabinett/Spätlese—Wehlener
Sonnenuhr by J.J. Prüm

Italy
Antinori Badia a Passignano
Chianti Classico

Antinori Chianti Classico Tenute
Marchese Riserva
Badia a Coltibuono Chianti
Classico Riserva
Col d'Orcia Brunello di Montalcino
Mastroberardino Taurasi
Melini Chianti Classico Riserva
Massovecchio
Ruffino Chianti Classico Riserva

New Zealand
Kim Crawford Chardonnay

Oregon
Argyle Pinot Noir

South Africa
Hamilton Russell Pinot Noir
and Chardonnay

Spain
Alvaro Palacios Priorat Les Terrasses
Bodega Montecillo Reserva
Bodegas Muga Reserva
C.U.N.E. Contino Reserva
La Rioja Alta Viña Ardanza
Pesquera

Washington State
Canoe Ridge Chardonnay and Merlot
Chateau Ste. Michelle Chardonnay
and Cabernet Sauvignon
Chateau Ste. Michelle Riesling Eroica
L'Ecole No. 41 Cabernet Sauvignon

Once-a-Year Wines
($$$$+)

It's easy to buy these kinds of wine when money is no object! Any wine
retailer would be more than happy to help you spend your money!

Creating an Exemplary Restaurant Wine List

WHEN I FIRST CONCEIVED the idea for this book in the early 1980s, I had two purposes in mind. One, to provide my wine students with an easier-to-understand text, and two, to satisfy a tremendous need in the hotel and culinary schools in the United States for a beginning guide to wine.

If you are a consumer, you may find this chapter to be interesting reading, since it was written for the young student who is about to embark on a career in the restaurant or hotel business.

Today the *Windows on the World Complete Wine Course* is used at top educational institutions such as the Culinary Institute of America, Cornell University, Michigan State University, University of Nevada (Las Vegas), Florida International University, and more than 100 other schools specializing in the hospitality industry.

Anyone who is even marginally aware of market trends knows that the popularity of wine has increased dramatically. Wine lists are no longer the sole province of an elite group of high-ticket, white-tablecloth culinary temples. There are ever-increasing ranks of customers who actively seek to enjoy wine in restaurants of all price levels.

To attract these potential customers, many a restaurateur has toyed with the idea of revamping and expanding his wine list. Yet, when confronted with the stark reality of such a task, many panic and accept the judgment of others, who have their own profit motives in mind. Once the restaurateur has acknowledged that it's time to start carrying more than house red, white, and rosé wines, he must lay the groundwork for building a list.

To illustrate: The hypothetical restaurant to which you've just taken the deed is a picturesque one-hundred-seat establishment located in a moderate-size city. Your restaurant is open for both lunch and dinner, and doesn't possess any definite ethnic identity, falling under that umbrella label of "Continental/American."

The following is a step-by-step procedure in question-and-answer format for building a wine list for your restaurant. Remember, the most important aspects of your wine list are that it should complement your menu offerings, be attractively priced, offer an appealing selection, and be easy for the customer to select from and understand.

What's your competition?

Before you consider which wineries to choose, or fret about whether to have 40 or 400 wines, take the time to investigate your market. Visit both the restaurants that attract the clientele you're aiming for (your target market) and the ones above and below your scale. Study how they merchandise wine, how well their staff serves it, and obtain a copy of their wine list—provided it's not chained to the sommelier's neck. Go during a busy dinner hour and observe how many bottles of wine are nestled in ice buckets or present on tables. Get a feeling for your competition's commitment to wine.

Can wine distributors offer help?

Yes. Contact the various suppliers in your area and explain to them what your objectives are. Ask them to suggest a hypothetical wine list—you're under no obligation to use it. Many wine distributors have specially trained people to work along with restaurateurs on their wine lists. They can also suggest ways of merchandising and promotion. Use them as a resource.

What's your storage capacity?

Wine requires specific storage conditions: an ideal temperature range of 55°F to 60°F, away from direct sunlight and excessive vibration. Your wine storage does not belong next to the dishwashing machine or the loading dock. How large a space do you have? Does it allow room for shelving? You'll want to store the bottles on their sides. How accessible is the wine to service personnel?

What are your consumers' preferences?

Preliminary research reveals that approximately 75 percent of the wine consumed in the United States is domestic and 25 percent is imported. The consumer preferences for our hypothetical restaurant's audience are predominantly American and French wines.

How long should your list be?

This is, in part, determined by storage space and capital investment. For our hypothetical restaurant, we decided to feature sixty wines on the list. Initially, this might seem like a lot for a 100-seat restaurant, but consider that those wines will be divided among sparkling, red, white, and rosé, and encompass the regions of France, the United States, Italy, Spain, Germany, Chile, and Australia. Sixty is almost the minimum number that will allow you the flexibility to offer a range of types and tastes and wine for special occasions, as well as for casual quaffing. We want our customers to know that wine matters in our restaurant.

What proportion of red to white?

In my wine school I always ask my students about their wine preferences. Ten years ago, the majority of my students preferred white wine, but today I find that my students have a decided preference for red wines. Because of this preference, I have decided to feature thirty-one reds and twenty-three whites, with six sparkling wines.

How should prices be set?

Don't plan on paying your mortgage with profits from your wine list! Pricing is dictated according to your selections. Aim for 60 percent of these wines to be moderately priced. Why? Because this is the price category in which the highest volume of sales will take place. For our purposes, mid-priced wines sell for between thirty and forty dollars. Therefore, we want thirty-six of our wines priced in this range. Of the remaining wines, 20 percent (twelve wines) would be priced less than thirty dollars, and 20 percent more than forty dollars.

The percentage of profit realized on wine is less than that realized on cocktails. However, the dollar-value profit is greater since the total sale is much more. Too many restaurateurs have intimidated much of the potential wine market by stocking only very rare, expensive wines and pricing them into the stratosphere. You want a wine list that will enable all your customers to enjoy a bottle (or two) of wine with their meal without having to float a bank loan. The bulk of your customers are looking for a good wine at a fair price— not a rare vintage wine at $400 a bottle.

What's your capital investment?

Determine with your accountant the amount of money you'll initially invest. Decide whether you want an inventory that will turn over in thirty to sixty days, or if you wish to make a long-term investment in cellaring wines. The majority of restaurants make short-term wine investments.

What will it actually cost?

Once we've decided on the number of wines on the list, and the general pricing structure, it's easy to determine what it will cost for one case of each wine. With some wines—the ones we anticipate will be very popular—our initial order will be for two cases, and for the more expensive wines, we'll start off with a half case. Our sixty-selection wine list will require an initial investment of approximately $10,200. Here's how we arrived at that "ballpark" figure:

LOW-PRICED WINES
12 cases @ $100 per case average $1200.00
MEDIUM-PRICED WINES
36 cases @ $150 per case average $5400.00
HIGH-PRICED WINES
12 cases @ $300 per case average $3600.00
TOTAL $10,200.00

What goes on the wine list?

There are many different styles of wine lists. Some opt for long descriptions of the wine's characteristics and feature facts and maps of viticultural regions.

For our restaurant, we're going to adopt a very straightforward approach.

• The list is divided into categories by type, and each type is divided into regions.

• The progression is: sparkling wine, white, red, and blush.

Each entry on the wine list should have the following information:

Bin number—This simplifies inventory and reordering, and assists both customer and staff with difficult pronunciation.

Name of wine—Be precise.

Vintage—This is often omitted on wine lists by restaurateurs who want to be able to substitute whatever is available. This practice is resented by anyone with a passing interest in wine, and with today's electronically-printed wine lists, it is, in my opinion, unacceptable.

Shipper—This information is very important for French wines, particularly those from Burgundy.

The type, style, and color of the paper you choose for your wine list are personal decisions. However, double-check all spelling and prices before the list is sent to the printer. It is very embarrassing when your customers point out spelling errors.

Which categories of wine should be used?

Our wine list includes six sparkling wines (three French, two American, and one Spanish). There are twenty-three white wines, thirty red wines, and one blush, for a grand total of sixty. On page 288 is one example of how the categories might be broken down. Of course, each restaurant should choose the wines according to availability and price.

Should I buy wine without tasting it?

Tasting the wines for the list is of utmost importance. If you've decided to feature a Chardonnay, contact your distributors and ask to taste all the Chardonnays that conform to your criteria of availability and price. Tasting the wines blind will help you make selections on the basis of quality rather than label. These tastings represent quite an investment of time. To choose a wine list of sixty wines, you could easily taste three times that many.

Once you've narrowed the field, try pairing the wines with your menu offerings. If possible, include your entire staff in these tastings, from managers to waiters to chefs. The more familiar they are with the wines on the list and the foods they complement, the better they'll be able to sell your selections.

Recheck the availability of your selected wines. Place your orders. Remember, you don't have to buy twenty-five cases of each wine.

Your restaurant's open—what's next?

This is a guideline to establishing an initial list. It is important to update and revise the list to meet the requirements of your customers and the ever-changing wine market.

Once your list has been implemented, it's imperative that you track wine

Building a Basic Restaurant Wine List

WHITE AND SPARKLING WINES (29)

FRENCH (8)

1 Mâcon Blanc
1 Chablis (Premier Cru)
1 Meursault
1 Puligny-Montrachet
1 Pouilly-Fuissé
1 Pouilly-Fumé or Sancerre
1 Alsace (1 Riesling or Gewürztraminer)
1 Bordeaux (1 Graves or Sauternes)

AMERICAN (12)

2 Sauvignon Blanc/Fumé Blanc
10 Chardonnay

ITALIAN (1)

1 Pinot Grigio

AUSTRALIAN (1)

1 Chardonnay

GERMAN (1)

1 Rhein or Mosel

SPARKLING (6)

3 French Champagne
  (2 Non-vintage, 1 Prestige Cuveé)
2 American Sparkling
  (2 different price categories)
1 Spanish Sparkling

RED AND BLUSH WINES (31)

FRENCH (7)

1 Beaujolais
2 Burgundy (such as Nuits-St-Georges, Pommard, Volnay)
3 Bordeaux (different price categories)
1 Rhône Valley (Côtes du Rhône, Châteauneuf-du-Pape, or Hermitage)

AMERICAN (18)

3 Merlot
3 Pinot Noir
2 Zinfandel
10 Cabernet Sauvignon

ITALIAN (2)

1 Chianti Classico Riserva
1 Barolo or Barbaresco

AUSTRALIAN (1)

1 Shiraz

SPANISH (1)

1 Rioja Reserva

CHILEAN (1)

1 Cabernet Sauvignon

BLUSH (1)

1 White Zinfandel

sales to determine how successful the list has been. Analyze your wine list with respect to the following factors:

- The number of bottles sold per customer (divide the number of bottles sold by the number of covers).
- Per-person check average for wine
- How much white wine to red (by percentage)—you might find that you need more or fewer whites, or more or fewer reds.
- The average price of a bottle of wine sold in the first three months.
- The ten most popular wines on the list.
- Instruct your staff to report any diner's request for a wine that's not on your list.

The steps involved in compiling our hypothetical list of sixty wines are the same steps used in compiling larger, more ambitious lists. Obviously, this list highlights only the major areas—sixty wines barely scratch the surface of what's available. True, with this size restriction we can't give great depth of selection, but the average customer, including myself, would still find something appealing.

THE PROGRESSIVE WINE LIST

In the mid-'80s at Windows on the World, we changed our wine list into a progressive wine list.

We decided Americans were more comfortable ordering wines by the specific grape varieties than by country and region. We featured the three major grape varieties for white wines—Riesling, Sauvignon Blanc/Fumé Blanc, and Chardonnay—and the three major varieties for reds—Cabernet Sauvignon, Pinot Noir, and Merlot. Then, we took all the wines and listed them by the major grape from which they are produced. For other wines made from other grape varietals, we created a separate heading on our wine list called "Worldly Wines."

We took it a step further and placed the wines on the list in order, from the lightest-style to the heaviest-style wines. This type of list is known as a "Progressive Wine List." There is a distinct advantage to this kind of wine list. By placing your lighter-bodied wines first and your heavier-bodied wines next, it helps your waitstaff to better recommend wines for the food the customer has chosen. To create a Progressive Wine List for your restaurant, you would lead off with the Rieslings, from the lightest- to the heaviest- style wines, then the Sauvignon Blancs/Fumé Blancs, and then the Chardonnays. For the red wines, you would begin with the Pinot Noirs, the Merlots, and then the Cabernet Sauvignons.

SEVERAL STUDIES have shown that wine by the glass represents more than 50% of wine sales in restaurants.

What about wine by the glass?

In my opinion, the most significant change in the twenty-four years since the first edition of this book has been the selling of wines by the glass. This is a home run for everyone—the consumer now has the opportunity to try different wines at different price points without buying a bottle. The waitstaff gets to learn more about different styles of wines by the glass than they do about wines listed on the wine list, plus they don't have to deal with the opening of a bottle since most of these wines are coming from the bar.

For the restaurant owner, you have a happier customer, a happier staff, faster service, and the potential to make more revenue and profits with an effective wine-by-the-glass program.

A note on pricing wines by the glass: The bottle of wine holds approximately twenty-five ounces; some restaurants get five five-ounce pours from each bottle and others get six four-ounce pours. My rule of thumb is to sell a glass for what the bottle costs me.

How many wines by the glass should I offer?

No matter how many wines you serve by the glass, whether it's only one or twenty, you must cork and store your wines (red and white) in the refrigerator overnight.

If you are interested in a wine-by-the-glass program, I suggest you start with six wines—three whites and three reds available at a low, medium, and high price point for each category. It is also a good idea (especially on your busy nights) to offer some very special, rare wines by the glass, just as a chef will add specials to his or her menus.

If you plan to have more than ten wines by the glass, invest in a wine preservation system. With less than ten, if you are selling an opened bottle of wine within forty-eight hours and you follow the overnight refrigeration procedure, you can probably do without a preservation system and still maintain the quality of the wine.

For further reading: *Restaurant Wine* newsletter by Ron Wiegand.

If you're a consumer, you now have an idea what's involved in creating a balanced wine list. For restaurateurs, the logistic of creating a wine list may seem Olympian. This chapter was offered to take some of the mystery out of this task.

Wine Service and Wine Storage in Restaurants and at Home

ongratulations! having reached this point, you've read about wine, and then how to combine wine with food, how to buy wine (and find the best values), and even how to create a restaurant wine list.

The last step in this process is serving the wine. Whether you are hosting a meal at a restaurant or at home, the goal of proper wine service should be to get the most enjoyment from your wine while eliminating any of the "performance anxieties" that would interfere with that enjoyment.

WINE IN RESTAURANTS

KEVIN ZRALY'S PET PEEVES OF RESTAURANT WINE SERVICE

- Not enough wine lists
- Incorrect information, i.e., wrong vintage and producer
- High markups
- Untrained staff
- Lack of corkscrews
- Out-of-stock wines
- Improper glassware
- Overchilled whites and warm reds

IN A RESTAURANT, I strongly recommend that you order all your wines at the beginning of the meal and have them opened so you're not sitting there without wine when your main course arrives. Even better, call the restaurant ahead of time and preorder your wines.

WINE SERVICE RITUAL

In most restaurants, the wine service ritual is as intimidating to the service staff as it is to the new wine consumer.

Picture this scenario: You've ordered your bottle of wine. The server now presents you with the bottle.

What do you do?

First, make sure that the wine presented to you at the table is the same wine you ordered from the wine list and is the correct vintage. Now the fun begins. The server opens the wine and presents you with the cork.

What are you supposed to do with the cork?

Nothing! After thirty years of being in the wine and restaurant industry, I have no idea why we give the customer the cork. In the movies, the bon vivant used to sniff the cork and wave it around ceremoniously. That looks good on film, but why would anyone want to smell a wet winey cork?!

If you really want to find out whether a wine is good, all you have to do is pour it in your glass, swirl it, and smell it. It's all in the smell!

The Magical Mystery Taste Test

Now you have to taste the wine. So many steps, so little time. While everyone at the dinner table is looking at you, the server pours the first taste of wine.

At this point I urge my students to remember that the first taste of wine is always a shock to your taste buds. If you are unsure, take a second taste to confirm your first impression. If you really want to impress your guests, simply smell the wine and nod your approval.

When should you send wine back in a restaurant?

Whenever you feel there is something wrong with the wine. Most restaurants will not have a problem with people sending a bottle of wine back, even though most of them sent back will be in good shape. For those of you who have heard about servers who recommend a wine to a table and add, "If you don't like it, I'll drink it myself," believe me, they do drink it, especially the good stuff!

What are the main reasons for sending wine back?

The first reason is that the wine is spoiled or oxidized. It has lost its fruit smell and taste, usually because of poor storage of the wine or the wine has passed its prime and should have been consumed earlier.

The second reason is that the wine is "corked." It is estimated that 3 to 5 percent of all the world's wines are corked, meaning that the cork that was used to seal the bottle was defective. You'll know immediately if a wine is corked because the smell of the wine is not of fruit, but is more like a dank, wet cellar or moldy newspaper.

DID YOU KNOW?

The wine-tasting ritual goes back to the days of kings, when royal tasters had to sample wine to make sure it did not contain poison before it was served to the king and his table. Unfortunately there was a high turnover rate of royal tasters!

SOME FACTORS that contribute to the making of a bad wine are:
- Faulty corks
- Poor selection of grapes
- Bad weather
- Bad winemaking
- High alcohol and tannin (bitterness)
- Herbaceousness
- Bacteria and yeast problems
- Unwanted fermentation in the bottle
- Hydrogen sulfide (rotten-egg smell)
- Excess sulfur dioxide
- Poor winery hygiene
- Unclean barrels
- Poor storage

AN OXIDIZED wine smells like Sherry or Madeira.

Is it necessary to spend a lot of money for wine in restaurants?

I've been in the restaurant business most of my life and also a customer in restaurants at least three times per week. So you might find it interesting that I never order a bottle of wine that costs more than seventy-five dollars (unless someone else is paying!). I do not believe that a restaurant is a great place for experimentation with wine because of the higher markup charged at most restaurants. We all know of great restaurants in our area that charge too much for wine and those that offer a fair markup. I find it more interesting to find a ten-dollar bottle of wine that tastes like a twenty-dollar bottle, a twenty-dollar bottle that tastes like a forty-dollar bottle, and so on. I go to the restaurants with the best wine prices. Restaurants with only high-priced wines and a limited choice of wines under seventy-five dollars do not get my credit card!

What is the tipping policy for sommeliers in restaurants?

As a sommelier (wine steward), I always appreciated receiving a gratuity for my knowledge and service. Although it should not be automatic, remember that "tips" (an acronym for "To Insure Proper Service") should definitely benefit your return to the restaurant.

A sommelier can make your dining experience better by:

- Not trying to sell you the most expensive wine on the list (unless, of course, you ask)
- Matching your food selection properly
- Decanting the wine
- Giving special wine suggestions that are not on the list
- Being attentive to your table
- Taking the label off the bottle (as a remembrance)
- Making your reservation and getting you a good table

Depending on service, I will tip anywhere from five to twenty dollars above the regular waiter gratuity for the sommelier.

TODAY'S ROYAL tasters are called sommeliers, which is the French word for wine steward. At the real ritzy restaurants, they are clad in tuxedos (a little bit more intimidation) and are usually seen wearing a chain with a little cup at the end of it. The cup is called a *tastevin* and is used by true sommeliers to taste the wine for you.

When everyone orders something different, what's the most versatile wine to complement dinner?

One of the biggest dilemmas my students confront is dining out with a large group and having to choose a wine to match a wide assortment of meat, fish, and vegetarian selections.

I usually order "safe" wines. My two favorite choices for a white wine are a Chardonnay without oak, such as a Mâcon from France or a slightly oak-aged Chardonnay from California or Australia.

As for the reds, my number one choice is definitely a Pinot Noir, most likely from California or Oregon, but if I can find a red Burgundy for under seventy-five dollars, that would also be one of my choices. Other red choices would be Brunello di Montalcino and Chianti Classico Riserva from Italy, and Reservas and Gran Reservas from Rioja, Spain.

These selections work well with meat, fish, and vegetarian choices.

WINE BY THE GLASS

In my opinion, the best thing that has happened to wine appreciation for both the consumer and the restaurateur is wine by the glass. If I'm going to do any experimentation in a restaurant, it will definitely be in a restaurant that has a good selection of wines by the glass. This also solves the age-old challenge of what wine to order when one person orders fish and another orders meat. You're no longer stuck with one bottle to accommodate everyone's different menu choices.

Another good thing about wine by the glass is that you get a hassle-free wine experience. That means:

1. You don't have to decipher the wine list. You don't even have to ask for it.
2. You don't have to approve the label.
3. You don't have to smell the cork.
4. You don't have to go through the tasting ritual.

THE AVERAGE temperature of a refrigerator is 38°F to 45°F.

TEMPERATURE

One of the biggest complaints in restaurant wine service is that the white wines are served too cold, but for me, being a red-wine drinker, my biggest complaint is that the red wines are served too warm.

White wines served too cold will hide the true character and flavor of the wine. Red wines served too warm change the balance of the components and hinder the true taste of the wine, increasing the sensation of alcohol and tannin over the fruit, especially in California wines.

In my own dining-out experience at least 50 percent of the time the wine is served at an improper temperature. It's easier to solve the problem for a white wine by leaving it in the glass and letting it warm up, but to chill down a red wine I have to order a bucket filled with ice and water and leave my wine in it for five to ten minutes to bring it to the correct temperature. I strongly recommend that you follow the same method, even though it may be a hassle. I can assure you that you'll enjoy your wines more.

KEVIN ZRALY'S QUICK TIPS ON TEMPERATURES FOR SERVING WINE

1. Great Chardonnays are best at warmer temperatures (55°F to 60°F) than whites made from Sauvignon Blanc or Riesling (45°F to 55°F).
2. Champagnes and sparkling wines taste better well chilled (45°F).
3. Lighter reds, such as Gamay, Pinot Noir, Tempranillo, and Sangiovese bring out a better balance of fruit to acid when served at a cooler temperature (55°F to 60°F) than wines such as Cabernet Sauvignon and Merlot (60°F to 65°F).

WINE AT HOME

STORAGE

I'm sure not everything you bought for your new wine cellar is going to be there five years from now, but some wines do improve with age, and you must protect your investment. The fun of wine collecting is trying wines at different stages in their growth.

Warmer temperatures prematurely age wine. How important is it to have proper storage for your new wine collectibles? Well, let me put it another way. Without proper storage, you will never know how good the wine could have been!

In wine collecting it seems as if it's always the cart before the horse. First I'll buy the wines, then I'll think about where I'm going to put them. Which of course is the wrong way to begin wine collecting. I've often asked my students how many of them live in châteaux and how many of them live in *apartamentos*? The *apartamento* people usually store their fine wines in the first door as you enter their *apartamento* next to their muddy shoes. If you are an apartment dweller and you have fine wine, you should look into wine storage in your area (or buy a château)!

While the optimum temperature is 55°F, I would rather the wine be a consistent 65°F all year long than swing in temperature from 55°F to 75°F. That is the worst scenario and can be very harmful to the wine. And just as warm temperatures will prematurely age the wine, temperatures too cold can freeze the wine, pushing out the cork and immediately ending the aging process.

When all else fails, especially for those living in a cozy apartment, put your wines in the refrigerator—both whites and reds—rather than risk storing them in warm conditions.

The second consideration in long-term wine storage (five years or more) is humidity. If the humidity is too low, your corks will dry out. If that happens, wine will seep out of the bottle. Again, if wine can get out, air can get in. Too much humidity, and you are likely to lose your labels. Personally, I'd rather lose my labels than lose my corks. My own wine cellar ranges from 55°F to 60°F and the humidity stays at a fairly constant 75 percent. Humidity and temperature are the most important things when it comes to wine storage, but I would also be careful of excessive vibration.

National statistics show that most wine purchased at retail stores will be consumed within three days of purchase, but for those of you who are collecting, you must protect your investment—whether you have a dozen bottles or two thousand.

Should my wines be stored horizontally or vertically?

Wine bottles should be stored horizontally, at up to a ten-degree angle, especially for long-term aging. The reason very simply is that when the bottle of wine is stored on its side, the wine will be in contact with the cork. This prevents any oxygen, the major enemy of wine, from getting into the bottle. If the wine is kept in a vertical position, the chances are great that the cork will dry out, the wine will evaporate, and the oxygen getting in will quickly spoil your wine.

P.S. Whoever wrote that it is important to turn the bottle of wine once a month was probably a beer drinker.

How long should I age my wine?

The Wall Street Journal recently came out with an article stating that most people have one or two wines that they've been saving for years for a special occasion. This is probably not a good idea!

More than 90 percent of all wine—red, white, and rosé—should be consumed within a year. With that in mind, the following is a guideline to aging wine from the best producers in the best years:

WHITE

California Chardonnay	3–8+ years
French White Burgundy	2–10+ years
German Riesling (Auslese, Beerenauslese, and Trockenbeerenauslese)	3–30+ years
French Sauternes	3–30+ years

RED

Bordeaux Châteaux	5–30+ years
California Cabernet Sauvignon	3–15+ years
Argentine Malbec	3–15+ years
Barolo and Barbaresco	5–25+ years
Brunello di Montalcino	3–15+ years
Chianti Classico Riservas	3–10+ years
Spanish Riojas (Gran Reservas)	5–20+ years
Hermitage/Shiraz	5–25+ years
California Zinfandel	5–15+ years
California Merlot	2–10+ years
California/Oregon Pinot Noirs	2–5+ years
French Red Burgundy	3–8+ years
Vintage Ports	10–40+ years

SOME WINES THAT ARE READY TO DRINK IMMEDIATELY:

Riesling (dry)
Sauvignon Blanc
Pinot Grigio
Beaujolais

There are always exceptions to the rules when it comes to generalizing about the aging of wine (especially considering the variations in vintages), hence the plus signs in the table above. I have had Bordeaux wines more than a hundred years old that were still going strong. It is also not unlikely to find a great

Sauternes or Port that still needs time to age after its fiftieth birthday. But the above age spans represent more than 95 percent of the wines in their categories.

P.S. The oldest bottle of wine still aging in Bordeaux is a 1797 Lafite-Rothschild.

What's the best way to chill a bottle of wine quickly?

The best technique that I have found for chilling white wines on short notice is to submerge the bottle in a mixture of ice, water, and salt. Within ten minutes the wine is ready for drinking.

THE CORKSCREW

THE CORKSCREW was patented in 1795 by an Englishman who obviously wanted to enjoy his wine without having to struggle to remove the cork.

How many of you have seen "other people" break a cork or even push the cork in the bottle when opening wine?

Many times this happens because people either use the wrong corkscrew or they open the wine incorrectly.

Of the many different kinds of corkscrews and cork-pullers available, the most efficient and easiest tool to use is the pocket model of the Screwpull, a patented device that includes a knife and a very long screw. Simply by turning in one continuous direction, the cork is extracted effortlessly. This is the best type of corkscrew for home use, and because it is gentle, it is best for removing long, fragile corks from older wines.

The corkscrew most commonly used in restaurants is the "waiter's corkscrew." Small and flat, it contains a knife, screw, and lever, all of which fold neatly into the handle.

For opening many bottles at a time, try the "rabbit" corkscrew.

How do I open a bottle of wine?

When opening a bottle of wine, the first step is to remove the capsule. You can accomplish this best by cutting around the neck on the underside of the bottle's lip. Once you remove the capsule, wipe the top of the cork clean—often dust or mold adheres to the cork while the wine is still at the winery, and before the capsule is put on the bottle. Next, insert the screw and turn it so that it goes as deeply as possible into the cork. Don't be afraid to go through the cork. I'd rather get a little cork in my wine than not get the cork out of the bottle.

The most important technique in opening a bottle of wine comes once you have lifted the cork one-quarter of the way out. You stop lifting and turn the screw farther into the cork. Now pull the cork three-quarters of the way out. This is the point where most people feel they're in control and start pulling and bending the cork. And, of course, they end up leaving a little bit of the cork still in the bottle. The best method at this point is to use your hand to wiggle out the cork.

Just to make you feel better: I still break at least a dozen corks a year.

TO DECANT, PERCHANCE TO BREATHE

Does a wine need to breathe?

This is one of the most controversial subjects among my wine friends, and everyone seems to have a different answer or angle regarding the question.

I think most of my peer group would agree that simply opening a bottle of wine an hour or two before service will not really help the wine. It also will not hurt the wine. It is probably a good idea if you are having a dinner party at home to open your wines before your guests arrive.

What is bothersome to me is when a waiter in a restaurant asks me whether I would like my wine to breathe before he or she serves it. A waiter once told me that the wine I had ordered needed at least thirty minutes' "breathing" before I could drink it. Not only do I disagree, but as a restaurant director, I certainly hope that the customer is ordering a second bottle thirty minutes later!

The major question still remains, though. Does a wine improve when it is taken out of the bottle and put into your decanter or glass? There are many schools of thought. I've had students swear to me that certain wines tasted much better after three hours in the decanter than when first served from the bottle. On the other hand, many studies with professional wine people have shown no discernible difference between most wines opened, poured, and consumed immediately and those that have been in a decanter over an extended period of time.

One thing for sure is that very old wine (more than twenty-five years) should be opened and consumed immediately. One of the most interesting wine experiences I ever had was early in my career, and involved a bottle of a

WHEN IN FRANCE, do as the French. My experience in visiting the French wine regions is that in Burgundy (Pinot Noir) they very rarely decant, but in Bordeaux (Cabernet Sauvignon and Merlot) they almost always decant.

MY PRIMARY reason for decanting a bottle of wine is to separate the wine from the sediment.

I HAVE had good and bad experiences with decanting. I must admit that some wines did get better, but many older wines, especially when exposed to air, actually deteriorated quickly. Words of wisdom: If you decide to decant an older bottle of wine, do it immediately before service, not hours ahead of time, to make sure that the wine will not lose its flavor by being exposed to the air for an extended period of time.

1945 Burgundy. When I opened the wine, the room filled with the smell of great wine. The first taste of the wine was magnificent. Unfortunately, fifteen minutes after opening the bottle, everything about the wine changed, especially the taste. The wine started losing its fruit, and the acidity overpowered the fruit.

What happened? Oxygen is the culprit here. Just as buried treasure is taken from the sea and kept in salt water to avoid exposing it to oxygen, wine is destroyed by exposure to oxygen. If I had decanted that wine first and left it to "breathe," I would never have had that first fifteen minutes of pleasure. This probably will not happen every time you open an old bottle of wine, but it is very important to be aware of how fragile older wines can be.

So what's my advice after opening thousands and thousands of bottles of wine? For me, Open it up, pour it in a glass, and enjoy the wine!

Which wines do I decant?

The three major wine collectibles are the ones that most likely will need to be decanted, especially as they get older and throw more sediment. The three major wine collectibles are:

1. Great châteaux of Bordeaux (ten years and older)
2. California Cabernets (eight years and older)
3. Vintage Port (ten years and older)

How do I decant a bottle of wine?

1. Completely remove the capsule from the neck of the bottle. This will enable you to see the wine clearly as it passes through the neck.
2. Light a candle. Most red wines are bottled in very dark green glass, making it difficult to see the wine pass through the neck of the bottle. A candle will give you the extra illumination you need and add a theatrical touch. A flashlight would do, but candles keep things simple.
3. Hold the decanter (a carafe or glass pitcher can also be used for this purpose) firmly in your hand.
4. Hold the wine bottle in your other hand, and gently pour the wine into the decanter while holding both over the candle at such an angle that you can see the wine pass through the neck of the bottle.

5. Continue pouring in one uninterrupted motion until you see the first signs of sediment.
6. Stop decanting once you see sediment. At this point, if there is still wine left, let it stand until the sediment settles. Then continue decanting.

GLASSWARE

Whether you're dining out or you're at home, the enjoyment of food and wine is enhanced by fine silver, china, linen, and, of course, glassware. The color of wine is as much a part of its pleasure and appeal as its bouquet and flavor. Glasses that alter or obscure the color of wine detract from the wine itself. The most suitable wineglasses are those of clear glass with a bowl large enough to allow for swirling.

A variety of shapes are available, and personal preferences should guide you when selecting glasses for home use. Some shapes, however, are better suited to certain wines than to others. For example, a smaller glass that closes in a bit at the top helps concentrate the bouquet of a white wine and also helps it keep its chill. Larger, balloon-shaped glasses are more appropriate for great red wines.

The most suitable Champagne glasses and the ones more and more restaurants are using are the tulip or the Champagne flute. These narrow glasses hold between four and eight ounces, and they allow the bubbles to rise from a single point. The tulip shape also helps to concentrate the bouquet.

How should I wash my wineglasses?

For your everyday drinking of wine, it's okay to wash your glasses in the dishwasher. But I've had many a great wine spoiled because of the detergent used. Therefore, my special wineglasses are not put into the dishwasher. They are washed by hand without any soap or detergent. Glasses are susceptible to scents, so mine are carefully dried hanging from a rack, not upside down on a counter or cloth.

THE BEST wineglasses are made by Riedel. The Riedel family of Austrian glassmakers has been on a crusade for more than 30 years to elevate wine drinking to a new level with their specially designed varietal glassware. They have designed their glasses to accentuate the best components of each grape variety. Riedel glassware comes in many different styles. The top of the line is the handcrafted Sommelier series. Next comes the Vinum. For your everyday meal—for those of you who do not want to spend a tremendous amount of money on glassware—they also produce the less expensive Overture series and the "O" wine tumbler for those of you who don't like stemware.

Frequently Asked Questions About Wine

What happens when I can't finish the whole bottle of wine?

This is one of the most frequently asked questions in Wine School (although I have never had this problem).

If you still have a portion of the wine left over, whether it be red or white, the bottle should be corked and immediately put into the refrigerator. Don't leave it out on your kitchen counter. Remember, bacteria grow in warm temperatures, and a 70°F+ kitchen will spoil wine very quickly. By refrigerating the wine, most wines will not lose their flavor over a forty-eight-hour period. (Some people swear that the wine even tastes better, although I'm not among them.)

Eventually, the wine will begin to oxidize. This is true of all table wines with an 8 to 14 percent alcohol content. Other wines, such as Ports and Sherries, with a higher alcohol content of 17 to 21 percent, will last longer, but I wouldn't suggest keeping them longer than two weeks.

Another way of preserving wine for an even longer period of time is to buy a small decanter that has a corked top and fill the decanter to the top with the wine. Or go to a hobby or craft store that also carries home winemaking equipment and buy some half bottles and corks.

Remember, the most harmful thing to wine is oxygen, and the less contact with oxygen, the longer the wine will last. That's why some wine collectors also use something called the Vacu-Vin, which pumps air out of the bottle. Other wine collectors spray the bottle with an inert gas such as nitrogen, which is odorless and tasteless, that preserves the wine from oxygen.

Remember, if all else fails, you'll still have a great cooking wine!

Why do I get a headache when I drink wine?

The simple answer may be overconsumption! Seriously though, more than 10 percent of my students are medical doctors, and none of them has been able to give me the definitive answer to this question.

Some people get headaches from white wine, others from red, but when it comes to alcohol consumption, dehydration certainly plays an important role in how you feel the next day. That's why for every glass of wine I consume, I will have two glasses of water to keep my body hydrated.

There are many factors that influence the way alcohol is metabolized in your system. The top three are:

1. Health
2. DNA
3. Gender

Research is increasingly leaning toward genetics as a reason for chronic headaches.

For those of you who have allergies, different levels of histamines are present in red wines; these can obviously cause discomfort and headaches. I myself am allergic to red wine and I "suffer" every day.

Many doctors have told me that food additives contribute to headaches. There is a natural compound in red wine called tyramine, which is said to dilate blood vessels. Further, many prescription medicines warn about combining with alcohol.

Regarding gender, due to certain stomach enzymes, women absorb more alcohol into their bloodstream than men do. A doctor who advises women that one glass of wine a day is a safe limit is likely to tell men that they can drink two glasses.

Do all wines need corks?

It is a time-honored tradition more than two centuries old to use corks to preserve wine. Most corks come from cork oak trees grown in Portugal and Spain.

The fact is that most wines could be sold without using cork as a stopper. Since 90 percent of all wine is meant to be consumed within one year, a screw cap will work just as well, if not better, than a cork for most wines.

Just think what this would mean to you—no need for a corkscrew, no broken corks, and, most important, no more tainted wine caused by contaminated cork.

I do believe that certain wines—those with potential to age for more than five years—are much better off using cork. But also keep in mind, for those real wine collectors, that a cork's life span is approximately twenty-five to thirty years, after which you'd better drink the wine or find somebody to recork it.

Some wineries now use a synthetic cork made from high-grade thermoplastic that is FDA-approved and also recyclable. These corks form a near-perfect

Screw-cap wines represent less than
5% of all bottled wine.

seal, so leakage, evaporation, and off flavors are virtually eliminated. They open with traditional corkscrews and allow wine to be stored upright.

But many wineries around the world use the Stelvin Screw Cap, especially in California (Bonny Doon, Sonoma Cutrer, etc.), Australia, and New Zealand.

What is a "corked" wine?

This is a very serious problem for wine lovers! There are some estimates that 3 to 5 percent of all wines have been contaminated and spoiled by a faulty cork. The principal cause of corked wine is a compound called TCA, short for 2,4,6-trichloranisole.

SOME WINERIES, especially in California, are now using synthetic corks to seal their wine. Since 1993, St. Francis Winery in Sonoma has sealed its wine with a synthetic cork. And Napa Valley's Plump Jack Winery released their $135/bottle 1997 Reserve Cabernet Sauvignon with a screw cap! Stay tuned for more.

When we find such a bottle at the Wine School, we make sure that every student gets a chance to smell a "corked" wine. It's a smell they won't soon forget!

Some of my students describe it as a dank, wet, moldy, cellar smell, and some describe it as a wet cardboard smell. It overpowers the fruit smell in the wine, making the wine undrinkable. It can happen in a ten-dollar bottle of wine or a thousand-dollar bottle of wine.

What's that funny-looking stuff attached to the bottom of my cork?

Tartaric acid, or tartrates, is sometimes found on the bottom of a bottle of wine or the cork. Tartaric acid is a harmless crystalline deposit that looks like glass or rock candy. In red wines, the crystals take on a rusty, reddish-brown color from the tannin.

Most tartrates are removed at the winery by lowering the temperature of the wine before it is bottled. Obviously this does not work with all wines, and if you keep your wine at a very cold temperature for a long period of time (for example, in your refrigerator), you can end up with this deposit on your cork.

Cool-climate regions like Germany have a greater chance of producing the crystallization effect.

Does the age of the vine affect the quality of the wine?

You will sometimes see on French wine labels the term *Vieilles Vignes* ("old vines"). In California, I've tasted many Zinfandels that were made from vines that were more than seventy-five years old. Many wine tasters, including myself, believe that these old vines create a different complexity and taste than do younger vines.

In many countries, grapes from vines three years old or younger cannot be made into a winery's top wine. In Bordeaux, France, Château Lafite-Rothschild produces a second wine, called Carruades de Lafite-Rothschild, which is made from the vineyard's youngest vines (less than fifteen years old).

As a vine gets older, especially over thirty years, it starts losing its fruit-production value. In commercial vineyards, vines will slow down their production at about twenty years of age, and most vines are replanted by their fiftieth birthday.

What are the hot areas in wine?

It seems as if most countries are catching the wine craze. Here are some areas where I have seen major growth and improvement in quality, especially with certain grape varieties, over the last twenty years:

New Zealand: Sauvignon Blanc, Chardonnay, and Pinot Noir
Chile: Cabernet Sauvignon
Argentina: Malbec
Hungary: Tokaji (one of the greatest dessert wines in the world)
Austria: Gruner Veltliner
Portugal: Not just Port anymore! Try Bacca Velha and you'll see what I mean.
South Africa: Sauvignon Blanc, Pinot Noir, and Syrah

And what will you be writing about in the year 2025?

Argentina, the United States, and Australia.

Besides the wine recommended in your book, can you recommend other wines from other countries?

Every year when I finish writing a new edition, people invariably ask me, "What did you add to the book this year to make it better?" My philosophy is, what did I learn this year that the beginning and intermediate wine person do not need to know; thus, every year my book gets more concise.

Of course, this is very unfair to great winemakers around the world whose countries or regions are not covered in my book. Here are some of the other great winemakers of the world:

WINES TO BUY NOW AND CELLAR FOR YOUR CHILD'S 21ST BIRTHDAY:

2005 Bordeaux, Sauternes, Burgundy, Southern Rhône, Piedmont, Tuscany, Germany, Rioja, Ribera Del Duero, Southern Australia, Napa Cabernet Sauvignon, Washington Cabernet Sauvignon
2004 Napa Cabernet Sauvignon, Piedmont
2003 Rhône (North and South). Sauternes, Bordeaux, Port
2002 Napa Cabernet Sauvignon, Germany*, Burgundy**, Sauternes
2001 Napa Cabernet Sauvignon, Sauternes, Germany*, Rioja and Ribera del Duero
2000 Bordeaux, Châteauneuf-du-Pape, Piedmont, Amarone, Port
1999 Piedmont, Rhône (North), California Zinfandel, Burgundy**
1998 Bordeaux (St-Émilion/Pomerol), Rhône (South), Piedmont (Barolo, Barbaresco)
1997 Napa Cabernet Sauvignon, Tuscany (Chianti, Brunello), Piedmont, Amarone, Port, Australian Shiraz
1996 Burgundy, Piedmont, Bordeaux (Médoc), Burgundy**, Germany*, vintage Champagne
1995 Bordeaux, Rhône, Rioja, Napa Cabernet Sauvignon, vintage Champagne
1994 Port, Napa Cabernet Sauvignon and Zinfandel, Rioja 1993 Napa Cabernet Sauvignon and Zinfandel
1992 Port, Napa Cabernet Sauvignon and Zinfandel
1991 Rhône (North), Port, Napa Cabernet Sauvignon
1990 Bordeaux, Napa Cabernet Sauvignon, Rhône, Burgundy**, Tuscany (Brunello), Piedmont, Amarone, Sauternes, Champagne, Germany*
1989 Bordeaux, Rhône, Piedmont, Rioja
1988 Sauternes, Rhône (North), Piedmont
1987 Napa Cabernet Sauvignon and Zinfandel
1986 Bordeaux, Sauternes, Napa Cabernet Sauvignon
1985 Bordeaux, Port, Rhône (North), Champagne, Piedmont, Amarone

*Auslese and above **Grand Cru

South Africa: Hamilton Russell, Kanonkop, Thelema, Glen Carlou
New Zealand: Cloudy Bay, Goldwater, Brancott, Morton Estate
Italy: Lungarotti, Mastroberardino, Jermann
Austria: Franz Prager, Kracher
Hungary: Royal Tokaji, Disznoko
Israel: Yarden
France: Mas de Daumas Gassac
Lebanon: Chateau Musar
Canada: Inniskillin

What are the most important books for your wine library?

Thank you for buying my wine book, which I hope you have found useful for a general understanding of wine. As with any hobby, there is always a thirst for more knowledge.

I hope that you noticed that at the end of each chapter, I recommended specific wine books for the different wine regions.

The following is a list of general books I consider required reading if you want to delve further into this fascinating subject:

The Essential Wine Book, Oz Clarke
Oz Clarke's New Encyclopedia of Wine
Oz Clarke's Wine Atlas
Great Wine Made Simple, Andrea Immer
Hugh Johnson's Modern Encyclopedia of Wine
World Atlas of Wine, Hugh Johnson
The Wine Bible, Karen MacNeil
Wine for Dummies, Ed McCarthy and Mary Ewing Mulligan
Keys to the Cellar by Peter D. Meltzer
Oxford Companion to Wine, Jancis Robinson
The New Sotheby's Wine Encyclopedia, Tom Stevenson

Since the above volumes are sometimes encyclopedic in nature, I always carry with me two pocket guides to wine:

Hugh Johnson's Pocket Encyclopedia of Wine
Oz Clarke's Pocket Wine Guide

Where can I get the best wine service in the United States?

The James Beard Awards have recognized the following restaurants with the Outstanding Wine Service Award:

1993	Charlie Trotter's, Chicago
1994	Valentino, Santa Monica
1995	Montrachet, New York
1996	Chanterelle, New York
1997	The Four Seasons, New York
1998	The Inn at Little Washington, Washington, Virginia
1999	Union Square Café, New York
2000	Rubicon, San Francisco
2001	French Laundry, Yountville, California
2002	Gramercy Tavern, New York
2003	Daniel, New York
2004	Babbo, New York
2005	Veritas, New York
2006	Aureole, Las Vegas
2007	Citronelle, Washington, D.C.

The past winners for Wine and Spirits Professional of the Year Award are:

1991	Robert Mondavi, Robert Mondavi Winery
1992	Andre Tchelistcheff, Beaulieu Winery
1993	Kevin Zraly, Windows on the World, New York
1994	Randall Grahm, Bonny Doon Vineyard, Santa Cruz
1995	Marvin Shanken, *Wine Spectator*
1996	Jack and Jaimie Davies, Schramsberg Vineyards
1997	Zelma Long, Simi Winery
1998	Robert M. Parker Jr., *The Wine Advocate*
1999	Frank Prial, *The New York Times*
2000	Kermit Lynch, Berkeley
2001	Gerald Asher, *Gourmet*
2002	Andrea Immer, French Culinary Institute
2003	Fritz Maytag, Anchor Brewing Co.
2004	Karen MacNeil, Culinary Institute of America
2005	Joseph Bastianich, Italian Wine Merchants, New York
2006	Daniel Johnnes, The Dinex Group, New York
2007	Paul Draper, Ridge Vineyards

What's the difference between California and French wines, and who makes the better wines?

You really think I'm going to answer that? California and France both make great wines, but the French make the best French wines!

From production strategy to weather, each region's profile is distinct. California wines and French wines share many similarities. The greatest similarity is that both France and California grow most of the same grape varieties. They also have many differences. The biggest differences are soil, climate, and tradition.

The French regard their soil with reverence and believe that the best wines only come from the greatest soil. When grapes were originally planted in California, the soil was not one of the major factors in determining which grapes were planted where. Over recent decades, this has become a much more important aspect for the vineyard owners in California, and it's not unheard of for a winemaker to say that his/her best Cabernet Sauvignon comes from a specific area.

As far as weather goes, the temperatures in Napa and Sonoma are different from those in Burgundy and Bordeaux. The fact is, that while European vintners get gray hair over pesky problems like cold snaps and rainstorms in the growing season, Californians can virtually count on abundant sunshine and warm temperatures.

Tradition is the biggest difference between the two, and I'm not just talking about winemaking. For example, vineyard and winery practices in Europe have remained virtually unchanged for generations; and these age-old techniques—some of which were written into law—define each region's own style. But in California, where few traditions exist, vintners are free to experiment with modern technology and create new products based on consumer demand. If you've ever had a wine called Two Buck Chuck, you know what I mean.

It is sometimes very difficult for me to sit in a tasting and compare a California Chardonnay and a French white Burgundy, since they have been making wines in Burgundy for the last 1,600 years and the renaissance of California wines is not yet 50 years old.

I buy both French and California wines for my personal cellar, and sometimes my choice has to do totally with how I feel that day or what food I'm having: Do I want to end up in Bordeaux or the Napa Valley?

Windows on the World:
A Personal History

BY KEVIN ZRALY

WINDOWS ON THE WORLD

WINDOWS ON THE WORLD OPENED its elevator doors for the first time on April 12, 1976, as a private luncheon club. The press had been writing about its opening for months, speculating on whether the whole project—both the restaurant and the World Trade Center complex that housed it—could be pulled off. After all, never before had a project this large been attempted. The World Trade Center was not only designed to include the tallest building in the world, it was going to be one of the largest urban centers ever built, with more than 40,000 World Trade Center office workers and 150,000 commuters passing through the complex every day.

All eyes were on Joe Baum, the man in charge of food services for the World Trade Center complex and mastermind behind the Windows on the World Restaurant. Joe had created some of New York's most successful landmark restaurants, including The Four Seasons, La Fonda del Sol, and the Forum of the Twelve Caesars.

He was known in the industry as a maverick, a restaurant genius, and a pit bull, and was often called all these names at the same time. In 1970, Joe Baum had signed a contract with the Port Authority of New York and New Jersey to design and manage all the restaurant and food-service areas of the World Trade Center. Joe and his associates, Michael Whiteman and Dennis Sweeney, conceived and organized twenty-two restaurants that were to be located throughout the complex.

Joe had grandiose ideas for the restaurant on top of the World Trade Center—ideas that would cost a lot of money.

Joe Baum on top of Two World Trade Center, overlooking One World Trade Center and Windows on the World

The Cellar in the Sky Restaurant

"As one of the first to join the 'Baum Squad,' I caught Joe's contagious fervor about Windows on the World right away and always knew it would be a smash success. And having the opportunity to work with Joe's 'kitchen cabinet'—James Beard, Craig Claiborne, Albert Stockli, Pierre Franey, Jacques Pépin, and Albert Kumin—was an education in itself. But nothing in my life will ever compare to opening day in April of 1976, a day of such excitement and wonder over this uber-restaurant."

—Dennis Sweeney

In the early '70s, New York City was in the midst of a severe fiscal crisis, and many New Yorkers were against building the World Trade Center to begin with. So Joe called his former classmate from Cornell, Curt Strand, president of Hilton International, to partner with the Port Authority, and together they formed a company called Inhilco.

Hilton International named Joe Baum president of Inhilco. With the backing of the Port Authority, more than $17 million was spent developing Windows on the World. Their main focus was the 107th floor of One World Trade Center, which they divided into five parts: The Restaurant (which seated nearly three hundred guests); The City Lights Bar; Hors D'Oeuvrerie, which served everything except the main course; The Cellar in the Sky, a glass-enclosed working wine cellar that seated just thirty-six guests and served a seven-course, five-wine, one-seating dinner; and six private banquet rooms capable of accommodating more than three hundred people. All together, Windows on the World spanned one acre, 107 stories up in the sky.

Joe Baum was a master contractor. He hired the best culinary talent available. He asked James Beard and Jacques Pépin to help develop the menus, Warren Platner to design the restaurant, Milton Glaser to design all the graphics, and Barbara Kafka to select everything from glassware to table settings. Joe also hired top restaurant managers, including Alan Lewis, his partner in other great New York restaurants, and a staff that would do anything Baum requested.

Joe was looking for a young American to run his wine department. I was lucky enough to be hired as the first cellarmaster at Windows on the World. I took the job after consulting with friends, many of whom warned me about leaving my job as a wine salesman to work at Windows. They gave me three reasons not to take the job:

1. In 1976, no one went downtown after 6 p.m.
2. Rooftop restaurants weren't considered quality operations.
3. Joe Baum had the reputation of being difficult to work with.

On all three counts, I learned it just wasn't so. In fact, I knew it was going to be a great restaurant and job when I asked Joe about creating the wine list.

He said: "It's very simple. I want you to create the biggest and the best wine list that New York has ever seen—and don't worry about how much it costs!" There I was, a twenty-five-year-old kid in a candy store—only it sold wine!

In May of 1976, before Windows on the World's official opening, the cover of *New York* magazine read: THE MOST SPEC-TACULAR RESTAURANT IN THE WORLD—HOW A BRILLIANT RESTAURA-TEUR CREATED A MASTERPIECE ON THE 107TH FLOOR OF THE WORLD TRADE CENTER. The article was written by the illustrious Gael Greene. Some of the superlatives from that article included: "a miracle," "a masterpiece," "a dream," "a triumph," and "almost unreal." It went on, "No other sky-high restaurant quite prepares you for the astonishment of the horizon."

The World Trade Center and Windows on the World also became symbols for the financial turnaround of New York City and their completion played a key role in the revitalization of lower Manhattan. This was underscored the year Windows on the World opened, 1976, which was also America's bicentennial year. Imagine an unobstructed view from the 107th floor of the spectacularly refurbished Statue of Liberty and the entire New York Harbor with its flotilla of tall ships. What a sight it was! Seeing the bicentennial fireworks from Windows became the hottest ticket in the world.

On that memorable July Fourth evening, I went alone to the top of One World Trade Center (the broadcast antennae and barriers had not yet been erected) and watched all the fireworks displays within a sixty-mile radius. I knew I'd made the right decision to work at Windows on the World; I remember thinking that life didn't get much better than this. I was serving wine to kings, queens, presidents, sports heroes, and movie stars. During the next five years, I met every celebrity I had ever heard of or read about.

Windows on the World was an instant success and was booked months in advance.

The Windows on the World Wine School has operated continuously for the last thirty years, since the opening of the restaurant in 1976, even during times of uncertainty. The Wine School started with a small group of ten lunch club members in 1976. Club members started inviting their friends, who then invited their friends. Soon the friends of club members outnumbered the club members. Still, the class list kept growing. In 1980, we opened the Wine

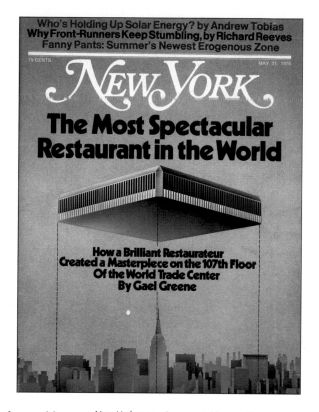

New York magazine cover, May 1976

"On July 4, 1976, Windows on the World was filled with celebrities for the bicentennial fireworks display. I was given the very pleasant task of escorting Princess Grace of Monaco. As we watched the extravaganza over the Statue of Liberty, Princess Grace held my hand very tightly because the fireworks made her somewhat nervous. I asked if the moment reminded her of To Catch a Thief and her very famous, very passionate scene with Cary Grant, set to the backdrop of fireworks. She was astounded I knew the film and the scene. So I told her that if she wanted to watch the film again, she could that night, because it was being shown at 11:30, on Channel 2."
—MELVIN FREEMAN, Page,
1976–93 and 1996–2001

School to the public. Since then, more than eighteen thousand students have attended classes.

So what made Windows on the World so great? Was it the sixty-second elevator ride? Was it the menu concept? Was it the youthful, energetic staff? Was it the extensive and outrageously low-priced wine list? Or was it the most spectacular view in the world?

For me, it was all of the above.

I continued as cellarmaster and, over the next four years, Windows on the World sold more wine than any other restaurant in the United States—and probably the world. The first five years of Windows on the World had been nonstop.

In 1980, I was named wine director for Windows on the World and Hilton International, and in 1981, I cofounded the New York Wine Experience, which is a celebration of the best wines from all over the world. The New York Wine Experience was held for the first three years at Windows on the World. Events like this, combined with our superb, well-priced wine list, helped establish Windows on the World as a true destination for wine lovers. We drew amateur wine enthusiasts as well as professionals; we attracted students and teachers alike; anyone seriously interested in wine stopped up for a bottle or two and a meal, including many who wanted to work in our wine cellar.

In 1985, my book, *Windows on the World Complete Wine Course*, made its debut, putting the Wine School in print for the first time. Over the next eight years, Windows continued to be one of the most successful restaurants in the world, the Wine School enrollment continued to grow, and the book became the best-selling wine book in the United States.

By the time Windows' fifteenth anniversary rolled around in 1991, both the restaurant and lower Manhattan had experienced their share of ups and downs. In 1987, Wall Street suffered a steep drop in stock market prices, which affected the restaurant business for several years. That same year, Ladbroke, a hotel and gaming company, bought out Hilton International and became the owner of Windows. As the stock market recovered, so did the businesses in lower Manhattan. By then, however, the notion of opening a fine restaurant near the financial district was no longer a novelty, and business at Windows faced quite a bit of competition from the proliferation of restaurants in its neighborhood.

Friday, February 26, 1993—The Bombing at the WTC

The first terrorist attack on the World Trade Center took place on February 26, 1993, at 12:18 p.m. Six people were killed, including one of our employees who worked in the receiving department in the World Trade Center basement. Food and wine author Andrea Immer, cellarmaster at the time, escorted all our patrons down 107 flights of stairs to safety.

Windows on the World shut down after the bombing, leaving more than four hundred food-service people without jobs. Within six months I was the last Windows on the World employee on the payroll. Windows on the World lay dormant from February 26, 1993, until June 1996. My Wine School coordinator and I were the only people allowed to enter Windows on the World after the 1993 bombing. It was a very lonely time.

Although the restaurant was closed from 1993 until 1996, with the help of Andrea Immer; Johannes Tromp, the director of Windows on the World; and Jules Roinnel, the director of The World Trade Center Club, the Wine School kept going. After the 1993 bombing, American Express gave us a temporary office across the street from the World Trade Center. Our view was of One World Trade Center and we would often stand at the windows watching tow trucks remove car after car—all destroyed—from the underground World Trade Center garages.

"I remember the first terrorist attack in 1993. We walked down the stairs with our lunch guests. I returned to work with Kevin when we reopened in 1996. I will never forget the extraordinary joy of seeing and touching those wines again, and welcoming our wine students back, and just being 'home,' where everything sparkled, most of all, the people."
—Andrea Immer, Cellarmaster, 1992–93

"Windows on the World was brilliantly conceived and truly unique. Being a Dutch immigrant managing the complex of restaurants and private dining rooms, with its 440 culinary and hospitality professionals high above Manhattan, was an unforgettable experience. I feel honored to have been a temporary guardian of this magnificent New York City institution."
—Johannes Tromp, Director, 1989–93

"Windows was silent when I ran the Wine School, but I relished its solitude. I roamed its floors, inspected its rooms, absorbed all that was left frozen in time. Like Jack Nicholson in The Shining, I could sense the energy it embodied, feel its buzz, envision its diners, find myself immersed in the dream. But unlike Jack's, my visions embodied warmth and peace. The space lived and breathed even when empty and will continue to do so in our hearts."
—Rebecca Chappa, Wine School Coordinator, 1994–95

"When I reflect on the 22 years I spent at Windows, no singular event immediately comes to mind but rather a series of images: the faces of children pressed up against the windows, the sun setting over the Statue of Liberty, and the laughter and joy of thousands of guests who created a lifetime of cherished memories."
—Jules Roinnel, Director, The World Trade Center Club, 1979–2001

World Trade Center, view from New Jersey

Both the Windows logo designed for the reopening in 1996 and the original logo on page 309 were created by Milton Glaser.

During the spring semester of 1993, Andrea Immer and I moved the school to the top of Seven World Trade Center. It remained there for the rest of 1993 and all of 1994. From 1994 to mid-1995, we held classes in the oval room on the forty-fourth floor of One World Trade Center, and from mid-1995 until the restaurant reopened in 1996, we operated from the newly reopened Vista Hotel (Marriott). I'm proud the Wine School remained open during those difficult years. It kept the memory of Windows alive and became a symbol to everyone that Windows would come back.

Following the 1993 bombing, the Port Authority concluded that both the 106th and 107th floors needed structural repairs, and meanwhile they would begin a search for a new operator for the restaurants. Again, requests for proposals went out, and this time more than thirty restaurant operators expressed interest in taking over Windows on the World. A review committee was formed by the Port Authority to examine each proposal and make recommendations. They narrowed it down to three entries: Alan Stillman of the Smith & Wollensky Restaurant Group, owners of Smith & Wollensky, the Post House, The Manhattan Ocean Club, Park Avenue Café, Cité, and Maloney & Porcelli; Warner LeRoy, of Tavern on the Green, Maxwell's Plum, and, later, the Russian Tea Room; and Joe Baum, the original creator of Windows on the World, who was then operating the Rainbow Room.

The Port Authority awarded the contract to Joe Baum, and the renaissance of Windows on the World began. Both Andrea Immer and I were brought back: Andrea to develop the wine list and beverage program, and I to continue with the Wine School. In all, the new staff totaled more than four hundred employees and represented some twenty-five nationalities.

The reopening of Windows on the World in June of 1996—twenty years almost to the month after its first opening—was accomplished with Joe Baum's usual kinetic energy, *joie de vivre*, and theatrical hoopla. Joe and his partners, the Emil family, reconceived the restaurant to make it the ultimate American-style food and wine experience. They took a big risk by committing to American cooking, but it paid off. Their willingness to do so made Windows the best it ever was. Windows on the World received two stars from *The New York Times*, three stars from *Crain's*, a 22 in the *Zagat* restaurant guide, and ranked in the top listings in *Wine Spectator* for overall dining.

Sadly, as Windows resumed its place on top of the world—having achieved high marks for both quality and service—Joe Baum died. That was in October 1998.

Over the next three years, Windows on the World remained a premier destination for wine, dining, and special events. One important change for me was the closing of The Cellar in the Sky, that intimate, romantically lit dining room lined with wine bottles. The Cellar in the Sky had been an integral part of the old Windows on the World, but by 1996 restaurants around the country were also doing food and wine pairings. It was time to replace The Cellar with something new. That something new was a restaurant called Wild Blue. Wild Blue was a restaurant within a restaurant—a place for chef Michael Lomonaco and his

Executive Chef/Director Michael Lomonaco and Chef Michael Ammirati in the Windows on the World kitchen

"There I saw the world from one and all other Windows of opportunity. . . . There my future began, from educating myself to educating my kids. There I learned the art of hospitality and traveled to teach others, those in the industry who remain and those who will be remembered."
—CARLOS A. GARCIA, Chief Executive Steward, 1976–86 and 1996–2001

"Windows on the World was an opportunity for Joe Baum, me, and everyone who worked with us to explore their dreams and pursue their ambitions in a spectacular environment."
—DAVID EMIL, Owner, 1993–2001

"Directing what was perhaps the most famous restaurant operation in the world challenged my intellectual and emotional being every day. The greatest of these challenges was, by the single element of location, that we would have the privilege of serving the most diverse clientele of any restaurant in the world and, in turn, employing the most diverse workforce of any restaurant in the world. The social interaction and collaboration of so many unique human beings, on both sides of the equation, was the Windows on the World experience. To have had the opportunity to be an integral part of the success of that collaboration made it an incredibly rewarding experience. In this respect, we truly were the Windows on the World." —GLENN VOGT, General Manager, 1997–2001

"Windows on the World was an inviting and hospitable place. . . . At Windows, everything seemed possible: fine wine, great food, and conviviality. There, floating high above the Earth, we never forgot we were citizens of the World. We welcomed one and all to dine, drink, and enjoy the sweetness of life."
—MICHAEL LOMONACO, Executive Chef/Director, 1997–2001

Overlooking the Brooklyn Bridge

"On Monday, September 10, I was hosting my class called 'Spirits in the Skybox,' a name that has haunting implications. When we finished, my cohost and I invited a few friends and some members of the press to join us for a quick drink in the bar. I usually have cocktails, but this time we ordered Champagne. There was no particular celebration; the group just seemed to click, so we had several more bottles of Champagne and food. A woman DJ began spinning records. Someone in our group knew her, so we stayed and ended the evening dancing. I awoke Tuesday morning to the horror of the terrorist attack that finished off that medium-sized city called the World Trade Center. I will cherish the gift I was given of that last spontaneous celebration in Joe Baum's majestic Windows on the World. We were unknowingly lifting our glasses that night in farewell to all those friends and colleagues we would lose the following day."

—DALE DEGROFF, Master Mixologist

staff to show off their culinary skills. Wild Blue received four stars from *Crain's*, a 25 in *Zagat*, a spot in the top-ten list of New York restaurants in *Wine Spectator*, and a rating as one of New York's best by *Esquire*.

Windows was the best it could be on September 10, 2001. It was generating more than $37 million in revenues and was the number one dollar-volume restaurant in the United States. Of that $37 million, upwards of $6 million came from the sale of wines from among our 1,400 different selections. We were enthusiastic about our future and were excitedly preparing for our twenty-fifth anniversary celebration in October 2001.

On September 11, 2001, the world changed. What began as a beautiful, pristine September morning ended in a dark nightmare of death and destruction. Seventy-two coworkers, one security officer, and six construction workers, who were building the new wine cellar at Windows on the World, died in the worst terrorist attack in American history.

For me, the day is still incomprehensible. The loss of my friends and coworkers remains heavy in my heart. The World Trade Center complex was my New York City. It was my neighborhood. I shopped there. I stayed at the Marriott Hotel with my family. I lost my home and community of twenty-five years, a community I watched being built.

Windows on the World no longer exists, but the reflections of those windows will remain with me forever.

Afterword:
Looking Back, with Gratitude

I WILL ALWAYS REMEMBER:

- working with and learning from John Novi at the Depuy Canal House, 1970–76

- my first visit to a winery, in 1970

- my first wine classes, in 1971 (one where I was a student and the other where I was the teacher)

- hitchhiking to California to visit the wine country, in 1972

- teaching a two-credit course as a junior in college (open only to seniors), in 1973

- Father Sam Matarazzo, my early spiritual leader

- living and studying wine in Europe, 1974–75

- planting my own vineyard (three-time failure), 1974, 1981, 1992; making my own wine, in 1984 (so-so)—trying again (2006)

- the excitement of opening Windows on the World, in 1976

- the support and friendship of Jules Roinnel, going back to our earliest days together at Windows

- wine tastings with Alexis Bespaloff

- to Mohonk Mountain House in New Paltz, New York, where ideas come easy

- Curt Strand and Toni Aigner of Hilton International

- evening, late-night, and early-morning wine discussions with Alexis Lichine in Bordeaux

- the adviser and great listener Peter Sichel, whose generosity of spirit inspired the way I teach and share my wine knowledge

- sharing great old vintages with Peter Bienstock

- Jules Epstein, for his advice and sharing his wine collection

- touring the world of Bordeaux with wine expert Robin Kelley O'Connor

- those who are no longer here to share a glass of wine: Craig Claiborne, Joseph Baum, Alan Lewis, Raymond Wellington, and my father, Charles

- creating and directing the New York Wine Experience, 1981–91

- witnessing the success of Michael Skurnik, who worked with me at Windows in the late 1970s and quickly rose to fame as a great importer of wines

- watching my former student Andrea Immer turn into a superstar wine-and-food personality and author

- the Food Network's *Wines A to Z* with Alan Richman
- reading and enjoying the observations of the great wine writers and tasters (listed throughout the book)
- having the opportunity to meet all the passionate winemakers, vineyardists, and owners of the great wineries of the world
- the wine events, wine dinners, and tastings around the country that I have had the privilege of attending
- all the groups that have invited me to entertain and educate them about wine
- writing the first chapter of this book with Kathleen Talbert in 1983
- the original Sterling Publishing team of Burton Hobson and Lincoln Boehm
- the Sterling Publishing team of Charles Nurnberg, Leigh Ann Ambrosi, and Chris Vaccari
- Steve Magnuson, who has a special talent for taking my ideas and helping me put them into words
- all my editors over the last two decades, especially Felicia Sherbert, Stephen Topping, Hannah Reich, Becky Maines, and Mary Hern
- the Sterling team assembled for the anniversary edition—Laurie Kahn, Rena Kornbluh, Julie Schroeder, Jeff Ward, Becky Maines, Mike Hollitscher, Pip Tannenbaum, and Sara Cheney—and the more recent additions of Chrissy Kwasnik, Nancy Field, Melanie Gold, Mary Hern, and Amy Lapides
- Karen Nelson for twenty-four years of beautiful cover designs
- Jim Anderson, who designed the original edition; and Richard Oriolo for capturing my spirit in subsequent editions
- Barnes & Noble, for always supporting my book
- Carmen Bissell, Raymond DePaul, and Faye Friedman for their help with the Wine School
- all my pourers at the school over the last thirty years
- having a great relationship with my New York City wine-school peers, especially Harriet Lembeck (Beverage Program); Mary Ewing-Mulligan (International Wine Center); and Robert Millman and Howard Kaplan (Executive Wine Seminars)
- the eighteen thousand students who have attended the Windows on the World Wine School, which celebrates its thirty-second anniversary in 2008
- the Baum-Emil team, who re-created Windows on the World in 1996
- conducting the Sherry-Lehmann/Kevin Zraly Master Wine Class with Michael Aaron, Michael Yurch, Chris Adams, and Shyda Gilmer
- Michael Stengel, Kathleen Duffy, Sarah Goldszak and all those nice people at the front desk (especially Tineka) at the Marriott Marquis Hotel NYC

- Jennifer Redmond, who assisted with the Wine School both in New Paltz and in New York City

- John and Linda Bono of Headington Wines & Liquors in New York City

- Alan Stillman, founder, chairman, and CEO of the Smith & Wollensky Restaurant Group

- the changing role of women in the wine industry (thank God!)

- being honored for "loving wine" by the James Beard Foundation

- teaching wine at Cornell University and the Culinary Institute of America

- being a member of the Culinary Institute's Board of Trustees

- Frank Prial and Florence Fabricant of *The New York Times* for their continual support

- Robert M. Parker Jr., who so generously donated his time and talents to aid the families of September 11th

- all the "special" wine friends who have "helped" me deplete my wine cellar over the years

- those who have tried to keep me organized in my business life: Ellen Kerr, Claire Josephs, Lois Arrighi, Sara Hutton, Andrea Immer, Dawn Lamendola, Catherine Fallis, Rebecca Chapa, Gina D'Angelo-Mullen, and Michelle Woodruff

- Ann Kiely for paying the bills on time

- Herb Schutte, my distributor analyst

- my wife of 18 years, Ana Fabiano (a former student who needed extra help after class!)

- my four best vintages: Anthony (1991), Nicolas (1993), Harrison (1997), and Adriana (1999)

- my mom, Kathleen

- my sisters, Sharon and Kathy

- everyone who worked at Windows on the World, especially my colleagues in the wine department

- my continuing grief at the loss of those friends and coworkers who lost their lives on September 11th

A Final Note

If this were an award-acceptance speech, I probably would have gotten the hook after the first three bullet points. Still, I'm sure I've forgotten to name at least one or two folks— an occupational hazard of consuming so much wine! So, to everyone I've ever known, from grammar school on: *May all your vintages be great!*

Glossary and Pronunciation Key

Acid: One of the four tastes of wine. It is sometimes described as sour, acidic, or tart and can be found on the sides of the tongue and mouth.

Aloxe Corton (ah-LOHSS cor-TAWN): A village in the Côte de Beaune in Burgundy, France.

Alsace (al-sas): A major white wine–producing region in northeastern France.

Amarone (ah-ma-ROH-nay): An Italian wine from Veneto made by a special process in which grapes are harvested late and allowed to "raisinate," thus producing a higher alcohol percentage in the wine and sometimes a sweet taste on the palate.

Amontillado (ah-mone-tee-YAH-doe): A type of Sherry.

AOC: Abbreviation for Appellation d'Origine Contrôlée; the French government agency that controls wine production there.

AP number: The official testing number displayed on a German wine label that shows the wine was tasted and passed government quality-control standards.

Aroma: The smell of the grapes in a wine.

Auslese (OUSE-lay-zeh): A sweet white German wine made from selected bunches of late-picked grapes.

AVA: Abbreviation for American Viticultural Area.

Barbaresco (bar-bar-ESS-coh): A full-bodied, DOCG red wine from Piedmont, Italy; made from the Nebbiolo grape.

Barbera (bar-BEAR-ah): A red grape grown in Piedmont, Italy, and California.

Barolo (bar-OH-lo): A full-bodied DOCG red wine from Piedmont, Italy; made from the Nebbiolo grape.

Beaujolais (bo-zho-LAY): A light, fruity red Burgundy wine from the region of Beaujolais; in terms of quality, the basic Beaujolais.

Beaujolais Nouveau (bo-zho-LAY noo-VOH): The "new" Beaujolais that's produced and delivered to retailers in a matter of weeks after the harvest.

Beaujolais-Villages (bo-zho-LAY vil-lahj): A Beaujolais wine that comes from a blend of grapes from designated villages in the region; it's a step up in quality from regular Beaujolais.

Beaune (BONE): French city located in the center of the Côte d'Or in Burgundy.

Beerenauslese (bear-en-OUSE-lay-zeh): A full-bodied, sweet white German wine made from the rich, ripe grapes affected by "botrytis."

Bitter: One of the four tastes.

Blanc de Blancs (blahnk duh blahnk): A white wine made from white grapes.

Blanc de Noir (blahnk duh nwahr): A white wine made from red grapes.

Botrytis cinerea (boh-TRY-tiss sin-eh-RAY-ah): A mold that forms on the grapes, known also as "noble rot," which is necessary to make Sauternes and the rich German wines *Beerenauslese* and *Trockenbeerenauslese*.

Bouquet: The smell of the wine.

Brix (bricks): A scale that measures the sugar level of the unfermented grape juice (*must*).

Brunello di Montalcino (brew-NELL-oh dee mon-tahl-CHEE-no): A high-quality DOCG red Italian wine from the Tuscany region.

Brut (BRUTE): The driest style of Champagne.

Cabernet Franc (cah-burr-NAY frahnk): A red grape of the Bordeaux region and the Loire Valley of France.

Cabernet Sauvignon (cah-burr-NAY so-vee-NYOH): The most important red grape grown in the world, yielding many of the great wines of Bordeaux and California.

Carmenre: A grape grown in Chile, sometimes mistaken for Merlot.

Chablis (shah-BLEE): The northernmost region in Burgundy; a wine that comes from Chardonnay grapes grown anywhere in the Chablis district.

Chambolle-Musigny (shahm-BOWL moos-een-YEE): A village in the Côte de Nuits in Burgundy, France.

Champagne: The region in France that produces the only sparkling wine that can be authentically called Champagne.

Chaptalization: The addition of sugar to the must (fresh grape juice) before fermentation.

Chardonnay (shahr-dun-NAY): The most important and expensive white grape, now grown all over the world; nearly all French white Burgundy wines are made from 100 percent Chardonnay.

Chassagne-Montrachet (shahs-SAHN-ya mown-rah-shay): A village in the Côte de Beaune in Burgundy, France.

Château (shah-TOH): The French "legal" definition is a house attached to a vineyard having a specific number of acres with winemaking and storage facilities on the property.

Château wine: Usually the best-quality Bordeaux wine.

Châteauneuf-du-Pape (shah-toh-nuff-dew-POP): A red wine from the southern Rhône Valley region of France; the name means "new castle of the Pope."

Chenin Blanc (SHEH-nin blahnk): A white grape grown in the Loire Valley region of France and in California.

Chianti (kee-AHN-tee): A DOCG red wine from the Tuscany region of Italy.

Chianti Classico (kee-AHN-tee class-ee-ko): One step above Chianti in terms of quality, this wine is from an inner district of Chianti.

Chianti Classico Riserva (key-AHN-tee class-ee-ko re-ser-va): The best-quality level of Italian Chianti, which requires more aging than Chianti and Chianti Classico.

Cinsault (san-SO): A red grape from France's Rhône Valley.

Classified châteaux: The châteaux in the Bordeaux region of France that are known to produce the best wine.

Concord: A red grape used to make some American and Eastern states' wines.

Colheita (coal-AY-ta): The term meaning "vintage" in Portuguese.

Cosecha (coh-SAY-cha): The term meaning "harvest" in Spanish.

Côte de Beaune (coat duh BONE): The southern portion of the Côte d'Or in Burgundy, known especially for fine white wines.

Côte de Nuits (coat duh NWEE): The northern portion of the Côte d'Or in Burgundy, known especially for fine red wines.

Côte d'Or (coat DOOR): The district in Burgundy that is known for some of the finest wines in the world.

Côte Rôtie (coat roe-TEE): A red wine from the northern Rhône Valley region of France.

Côtes-du-Rhône (coat dew ROAN): The Rhône Valley region of France; also the regional wine from this district.

Cream Sherry: A type of Sherry made from a mixture of Pedro Ximénez and Oloroso.

Crianza (cree-AHN-za): A wine aged a year in oak and a year in the bottle. It is the most basic and least expensive quality level of Rioja wine.

Crozes-Hermitage (crows air-mee-TAHZH): A red wine from the northern Rhône Valley region of France.

Cru Beaujolais (crew bo-zho-LAY): The top grade of Beaujolais wine, coming from any one of ten designated villages in that region of France.

Cru Bourgeois (crew bour-ZHWAH): A list of 247 châteaux in Bordeaux that have been recognized for their quality.

Decanting: The process of pouring wine from its bottle into a carafe to separate the sediment from the wine.

Dégorgement (day-gorzh-MOWN): One step of the Champagne method used to expel the sediment from the bottle.

Demi-sec (deh-mee SECK): A Champagne containing a high level of residual sugar.

DOC: Abbreviation for Denominazione di Origine Controllata, the Italian government agency that controls wine production.

DOCG: Abbreviation for Denominazione di Origine Controllata e Garantita; the Italian government allows this marking to appear only on the finest wines. The G stands for "guaranteed."

Dolcetto (dohl-CHET-toh): A red wine from Piedmont, Italy, that is lighter in style than a Barolo or Barbaresco.

Dosage (doh-SAHZH): A combination of wine and cane sugar that is used in making Champagne.

Edelfäule (EH-del-foy-luh): A German name for the mold that forms on the grapevines when the conditions permit it. (See also *Botrytis cinerea* and "Noble Rot.")

Estate-bottled: Wine that's made, produced, and bottled by the vineyard's owner.

Extra dry: Less dry than brut Champagne.

Fermentation: The process by which grape juice is made into wine.

Fino (FEE-noh): A type of Sherry.

First growth: The highest-quality Bordeaux château wine from the Médoc Classification of 1855.

Flor: A type of yeast that develops in some Sherry production.

Fortified wine: A wine such as Port or Sherry that has additional grape brandy that raises the alcohol content.

Gamay (gah-MAY): A red grape used to make Beaujolais wine.

Garnacha (gar-NAH-cha): A red grape grown in Spain. It is the same as the Grenache grape grown in the Rhône Valley region of France.

Gevrey Chambertin (zhehv-RAY sham-burr-TAN): A village in the Côte de Nuits in Burgundy, France.

Gewürztraminer (geh-VERTZ-tra-MEE-ner): The "spicy" white grape grown in Alsace, California, and Germany.

Gran Reserva: A Spanish wine that's had extra aging.

Grand Cru (grawn crew): The highest classification for wines in Burgundy.

Grand Cru Classé (grawn crew clas-SAY): The highest level of the Bordeaux classification.

Graves (grahv): A region in Bordeaux producing dry red and white wines.

Grenache (greh-NOSH): A red grape of the Rhône Valley region of France.

Gutsabfüllung: A German word for an estate-bottled wine.

Hectare: A metric measure that equals 2.471 acres.

Hectoliter: A metric measure that equals 26.42 U.S. gallons.

Halbtrocken: The German term meaning "semidry."

Hermitage (air-mee-TAHZH): A red wine from the northern Rhône Valley region of France.

Jerez de la Frontera (hair-eth day la fron-TAIR-ah): One of the towns in Andalusia, Spain, where Sherry is produced.

Jug wine: A simple drinking wine from California that is sold in "jug" bottles.

Kabinett (kah-bee-NETT): A light, semi-dry German wine.

Liebfraumilch (LEEB-frow-milch): An easy-to-drink white German wine; it means "milk of the Blessed Mother."

Liqueur de Tirage (lee-KERR deh teer-AHZH): A blend of sugar and yeast added to Champagne to begin the wine's second fermentation.

Loire (LWAHR): A major wine-producing region in northwestern France.

Long-vatted: A term for a wine fermented with the grape skins for a long period of time to acquire a rich red color.

Mâcon Blanc (mac-CAW blahnk): The most basic white wine from the Mâconnais region of Burgundy, France.

Mâcon-Villages (mac-CAW vee-LAHZH): A white wine from designated villages in the Mâconnais region of France; a step above the Mâcon Blanc quality.

Malbec: The major red grape grown in Argentina. It is also grown in Bordeaux.

Manzanilla (mahn-than-NEE-ya): A type of Sherry.

Margaux (mar-GO): A village and district in the Bordeaux region in France.

Mechanical harvester: A machine used on flat vineyards. It shakes the vines to harvest the grapes.

Médoc (may-DOCK): A district in the Bordeaux region in France.

Merlot (mehr-LOW): The red grape grown primarily in the Bordeaux region of France and in California.

Méthode Champenoise (may-TUD shahm-pen-WAHZ): The method by which Champagne is made.

Meursault (mehr-SOH): A village in the Côte de Beaune in Burgundy, France.

Microclimate: A term that refers to an area that has a climate within a climate. While one area may be generally warm, it may contain a cooler *microclimate* or region.

Morey-St-Denis (mor-RAY san duh-NEE): A village in the Côte de Nuits in Burgundy, France.

Mosel-Saar-Ruwer (MO-z'l sahr roo-ver): A region in Germany that produces a light-style white wine.

Müller-Thurgau (MEW-lurr TURR-gow): A cross between the Riesling and the Silvaner grapes of Germany.

Muscadet (moos-cah-DAY): A light, dry wine from the Loire Valley of France.

Muscat Beaumes-de-Venise (mus-CAT bome deh ven-EASE): A sweet wine from the Rhône Valley region of France.

Must: Grape juice before fermentation.

Nebbiolo (nehb-bee-OH-loh): A red grape grown in Piedmont, Italy, which produces some of the finest Italian wine, such as Barolo and Barbaresco.

"Noble Rot": See *Botrytis cinerea*.

Non-vintage Champagne: Champagne made from a blend of vintages (more than one year's crop); it is more typical of the house style than vintage Champagne.

Nose: The term used to describe the bouquet and aroma of wine.

Nuits-St-Georges (nwee san ZHORZH): A village in the Côte de Nuits in Burgundy, France.

Official Classification of 1855: A classification drawn up by wine brokers of the best Bordeaux, Médoc, and Sauternes châteaus.

Pauillac (PAW-yak): A village and district in the Bordeaux region of France.

Pessac-Leognan (pes-sack lee in yawn): An inner district of the Graves region in Bordeaux, France, making both red and white wine.

Petite Sirah: A red grape grown primarily in California.

Pfalz (faults): A wine region in Germany.

Phylloxera (fill-LOCK-seh-rah): A root louse that kills grape vines.

Piedmont (peed-MON-tay): One of the most important wine districts in Italy.

Pinot Blanc: A white grape grown primarily in the Alsace region of France.

Pinot Grigio (PEE-noh GREE-jee-o): The most popular white wine from Italy made from the grape variety called Pinot Grigio, aka Pinot Gris in France and the United States

Pinot Meunier (PEE-noh muhn-YAY): A red grape grown primarily in the Champagne region of France.

Pinot Noir (PEE-noh nwahr): All red French Burgundy wines, except Beaujoulais, are made from 100 percent Pinot Noir grapes. It is also very successful in California and Oregon.

Pomerol (palm-muh-roll): A district in the Bordeaux region of France.

Pommard (poh-MAR): A village in the Côte de Beaune in Burgundy, France.

Pouilly-Fuissé (pooh-yee fwee-SAY): The highest-quality French white Mâconnais wine made from 100 percent Chardonnay.

Pouilly-Fumé (pooh-yee fooh-MAY): A dry white wine from the Loire Valley region of France made from Sauvignon Blanc.

Premier Cru: A wine that has special characteristics that comes from a specific designated vineyard in Burgundy, France, or is blended from several such vineyards.

Priorat: A major wine region in Spain.

Proprietary wine: A wine that's given a brand name like any other product and is marketed as such, e.g., Riunite, Mouton-Cadet.

Puligny-Montrachet (pooh-lean-yee mown-rah-SHAY): A village in the Côte de Beaune in Burgundy, France.

Qualitätswein (kval-ee-TATES-vine): A German term meaning "quality wine."

Qualitätswein mit Prädikat (kval-ee-TATES-vine mitt pray-dee-KAHT): The highest level of quality German wine.

Reserva/Riserva: A term that means a wine has extra aging; it is often found on Spanish, Portuguese, and Italian wine labels.

Reserve: A term sometimes found on American wine labels. Although it has no legal significance, it usually indicates a better-quality wine.

Residual sugar: An indication of how dry or sweet a wine is.

Rheingau (RHINE-gow): A region in Germany.

Rheinhessen (RHINE-hess-en): A region in Germany.

Rheinpfalz (RHINE-faults): A region in Germany. The official name has now been changed to Pfalz.

Ribera del Duero (ree-BAY-rah dell dway-roh): A winegrowing region in Spain.

Riddling: One step of the Champagne-making process in which the bottles are turned gradually each day until they are upside down, with the sediment resting in the neck of the bottle.

Riesling: A white grape grown in Alsace, France, and the United States.

Rioja (ree-OH-ha): A wine region in Spain.

Ruby Port: A dark and sweet fortified wine blended from non-vintage wines.

Sancerre (sahn-SEHR): A dry white wine from the Loire Valley region of France.

Sangiovese (san-jo-VAY-zay): A red grape grown primarily in Tuscany, Italy.

Sauternes (soh-TURN): A sweet white wine from the Bordeaux region of France.

Sauvignon Blanc (SOH-veen-yown blahnk): A white grape grown primarily in the Loire Valley, Graves, and Sauternes regions of France, in Washington State, New Zealand, and California (where the wine is sometimes called Fumé Blanc).

Sémillon (say-mee-YAW): A white grape found in the Graves and Sauternes regions of Bordeaux, France, and Australia.

Shiraz (SHEER-oz): A red grape grown primarily in Australia, aka Syrah.

Short-vatted: A term for a wine fermented with the grape skins for only a short time.

Solera system (so-LEHR-ah): A process used to blend various vintages of Sherry.

Sommelier (so-mel-YAY): The French term for cellarmaster, or wine steward.

Spätlese (SHPATE-lay-zuh): A white German wine made from grapes picked later than the normal harvest.

Stainless-steel tank: A container that (because of its ability to control temperature) is used to ferment and age some wines.

St-Émilion (sahnt ay-meel-YOHN): A district in the Bordeaux region of France.

St-Estèphe (sahnt ay-STEFF): A village and district in the Bordeaux region of France.

St-Julien (sahnt zhoo-lee-EHN): A village and district in the Bordeaux region of France.

St-Véran (sahn vay-RAHN): A white Mâconnais wine one step above Mâcon-Villages in quality.

Sulfur dioxide: A substance used in winemaking and grape growing as a preservative, an antioxidant, and also as a sterilizing agent.

Süssreserve: The unfermented grape juice added to German wine after fermentation to give the wine more sweetness.

Syrah (see-RAH): A red grape grown in the Rhône Valley region of France and Australia, aka Shiraz.

Tafelwein (taf'l VINE): A German table wine.

Tannin: A natural compound and preservative that comes from the skins, stems, and pips of the grapes and also from the wood in which wine is aged.

Tawny Port: A Port that is lighter, softer, and aged longer than Ruby Port.

TBA: Abbreviation for the German wine Trockenbeerenauslese.

Tempranillo (temp-rah-NEE-yoh): A red grape grown primarily in Spain.

Trebbiano (treb-bee-AH-no): A white grape grown in Italy.

Trockenbeerenauslese (troh-ken-bear-en-OUSE-lay-zuh): The richest and sweetest wine made in Germany from the most mature grapes.

Tuscany (TUSS-cah-nee): A region in Italy.

Varietal wine: A wine that is labeled with the predominant grape used to produce the wine, i.e., a wine made from Chardonnay grapes would be labeled "Chardonnay."

Veneto wines: A wine region in Italy producing Valpolicella, Bardolino, Soave, and Amarone.

Village wine: A wine that comes from a particular village in Burgundy.

Vino Nobile di Montepulciano (VEE-noh NOH-bee-leh dee mon-teh-pull-CHAH-noh): A DOCG red wine from the Tuscany region of Italy.

Vintage: The year the grapes are harvested.

Viognier (vee-own-YAY): A white grape from the Rhône Valley region of France and California.

Vitis labrusca (VEE-tiss la-BREW-skah): A native grape species in America.

Vitis vinifera (VEE-tiss vih-NIFF-er-ah): The European grape species used to make European and California wine.

Volnay (vohl-NAY): A village in the Côte de Beaune region of Burgundy, France.

Vosne Romanée (vohn roh-mah-NAY): A village in the Côte de Nuits region of Burgundy, France.

Vougeot (voo-ZHOH): A village in the Côte de Nuits region of Burgundy, France.

Vouvray (voo-VRAY): A white wine from the Loire Valley region of France; it can be dry, semi-sweet, or sweet.

Wood Port: Ruby or Tawny Port; they're ready to drink as soon as you buy them.

Zinfandel (zin-fan-DELL): A red grape grown in California.

Index

Note: Page references in **bold** indicate label(s) included on that page.

Carbonnier

Gladys Harinchi

Robin Kelley O'Connor

Alistair Robertson

(Louis) LATOUR

Marion R. Shenker

LF Bouchard

Rodney D. Strong

Pio Boffa

Stéfan m

James Trezise

Louis P. Martini

Alan Lewis

Mike Stephens

Ed Sbragia

John L. Meier

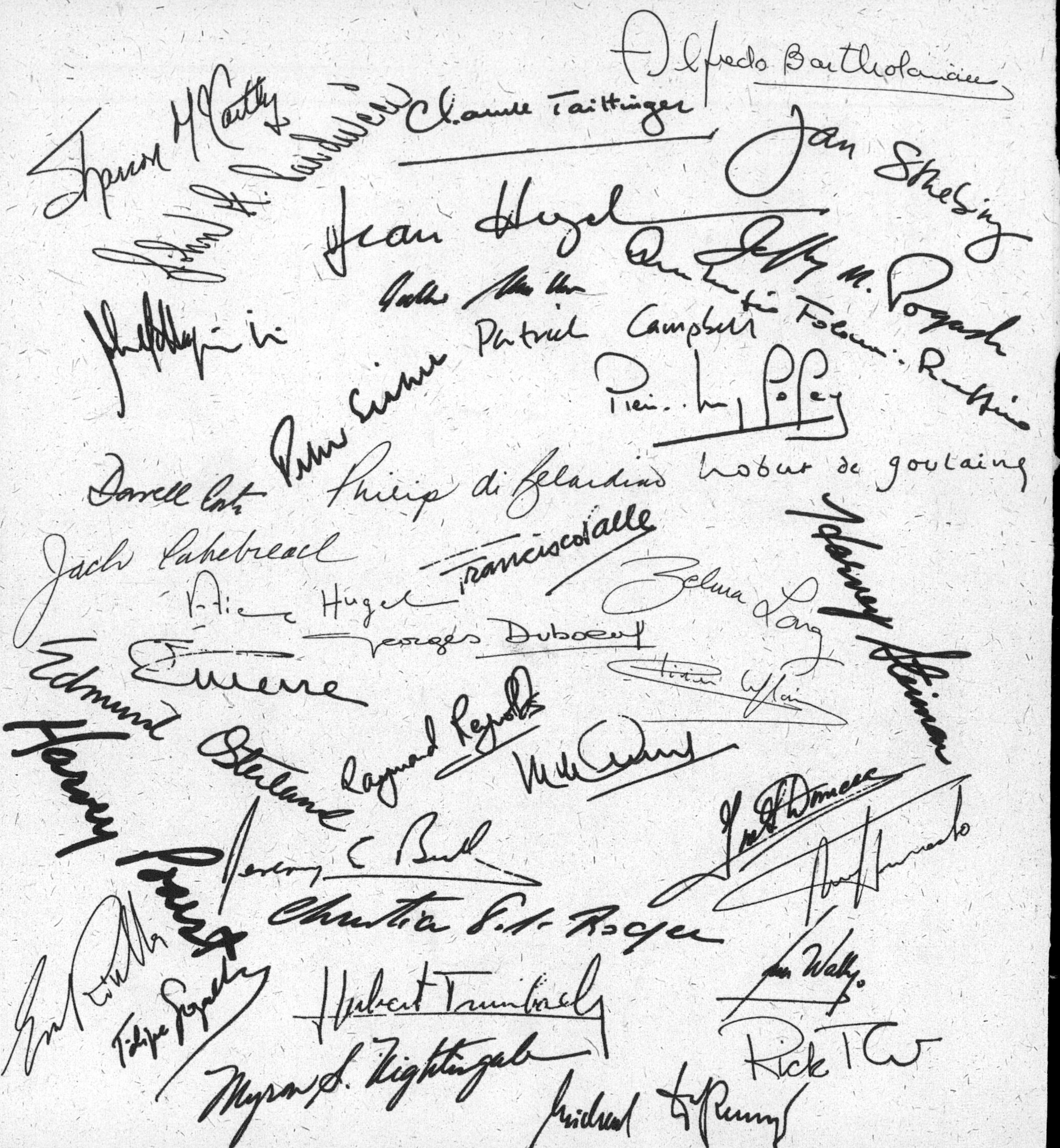
Claude Taittinger
Jean Hugel
Patrick Campbell
Robert de Goulaine
Philip di Belardino
Darrell Corti
Jack Cakebread
Franciscovalle
Zelma Long
Georges Duboeuf
Hubert Trimbach
Myron S. Nightingale